普通高等教育机械类系列教材

机械工程专业英语
（第4版）

唐一平 主 编
李 艳 徐海黎 副主编

电子工业出版社
Publishing House of Electronics Industry
北京·BEIJING

内 容 简 介

本书按照"内容新,知识面广,难度适中,反映最新科研成果"的原则编写而成,内容包括工程材料、铸造、锻造与模具、常规机械加工工艺、特种加工工艺、高速切削加工、公差与配合、定位与夹具、齿轮传动、直接传动技术、数控技术、数控加工中心、自动控制技术、CAD/CAM、传感器、机器人、增材制造、新能源、现代制造技术的发展、信息技术与制造业、制造与工程师的作用等。附录给出了科技论文英文摘要写作要点、麻省理工学院简介、机床说明书翻译样本及校园常用词汇等内容。

本书可供高等学校机械类及相关专业高年级本科生作为专业英语教材使用,也可作为企业和科研单位技术人员的专业英语阅读参考书。

未经许可,不得以任何方式复制或抄袭本书之部分或全部内容。
版权所有,侵权必究。

图书在版编目(CIP)数据

机械工程专业英语 / 唐一平主编. -- 4 版.
北京 : 电子工业出版社, 2024. 7. -- ISBN 978-7-121-48561-9

Ⅰ. TH

中国国家版本馆 CIP 数据核字第 2024RS7329 号

责任编辑:张天运
印　　刷:涿州市京南印刷厂
装　　订:涿州市京南印刷厂
出版发行:电子工业出版社
　　　　　北京市海淀区万寿路 173 信箱　邮编:100036
开　　本:787×1092　1/16　印张:15.75　字数:524.16 千字
版　　次:2009 年 10 月第 1 版
　　　　　2024 年 7 月第 4 版
印　　次:2024 年 7 月第 1 次印刷
定　　价:49.00 元

凡所购买电子工业出版社图书有缺损问题,请向购买书店调换。若书店售缺,请与本社发行部联系,联系及邮购电话:(010)88254888,88258888。

质量投诉请发邮件至 zlts@phei.com.cn,盗版侵权举报请发邮件至 dbqq@phei.com.cn。
本书咨询联系方式:zhaoys@phei.com.cn。

第 4 版前言

《机械工程专业英语》第 3 版自 2017 年 8 月出版发行以来得到各有关高校师生的关注与支持，目前采用该教材的院校已有四十多所，它们普遍认为该教材内容基本涵盖了机械专业本科生应掌握的有关机械制造领域的专业知识，也反映了机械制造行业近年来的发展成果。为进一步拓宽、更新、提升教材内容，并根据部分授课教师要求增加近年来在新能源和智能机器人领域所取得的研究成果的建议，本教材第 4 版新增了新能源单元，并在原 18 单元后面增加了智能机器人的有关内容。原 11 单元液压传动由于内容陈旧，予以删除。

编者想再一次强调，后面的附录部分包括了教材部分单元的中文译文、科技论文英文摘要写作要点、世界著名工科院校简介、机械设备说明书翻译样本及校园常用词汇等内容。这些内容对于学生的英文写作水平的提高十分重要，建议授课教师根据各校各专业具体情况有选择性地讲授其中某些内容，也希望读者特别留意这部分内容。

采用本书的授课教师提供有关院校的教学计划和任务书等信息，即可索取本教材的参考译文及电子课件（编者邮箱：yptang@mail.xjtu.edu.cn，出版社联系方式：zhaoys@phei.com.cn）。

编　者

2024 年 6 月

作 者 简 介

唐一平，男，湖南东安人。西安交通大学教授，博士生导师，中国民主同盟盟员。1968 年毕业于西安交通大学机械制造系，于 1979 年回校读研，1982 年获西安交通大学固体力学专业工学硕士学位。1983 年 1 月—1995 年 9 月在西安理工大学机械工程系任讲师、副教授。1995 年 10 月调入西安交通大学机械工程学院工作，2002 年获评教授职称。曾主持或承担多项国家自然科学基金及其他国家和省部级科研项目，并长期负责全校研究生课程"实用科技英文写作"的授课；先后获国家科学技术进步奖二等奖 1 项，省部级科技成果奖 2 项，西安交通大学科研和教学成果奖 6 项；2008 年被宝钢教育基金会授予宝钢优秀教师奖，2011 年被民盟中央委员会授予"先进个人"称号。近年来在国内外知名杂志发表论文并被 SCI 收录 80 余篇，主编出版英文版教材 5 本，并在世界图书出版公司出版了我国第一部高等教育双语工具书《中国高等教育汉英词典》。

CONTENTS

Unit 1　Engineering Materials (Ⅰ) ·· 1
　1.1　Introduction ·· 1
　1.2　Ferrous Metals and Alloys ··· 2
　1.3　Nonferrous Metals and Alloys ··· 4
　Notes ··· 7
　Glossary ·· 9

Unit 2　Engineering Materials (Ⅱ) ·· 10
　2.1　Introduction ·· 10
　2.2　Ceramics ··· 10
　2.3　Polymers ··· 12
　2.4　Composite Materials ·· 14
　2.5　Metamaterials ··· 15
　Notes ··· 16
　Glossary ·· 17

Unit 3　Casting ·· 20
　3.1　Introduction ·· 20
　3.2　Sand Casting ··· 20
　3.3　Investment Casting ··· 21
　3.4　Expendable-pattern Casting ··· 22
　3.5　Centrifugal Casting ·· 23
　3.6　Inspection of Casting ··· 24
　Notes ··· 24
　Glossary ·· 25

Unit 4　Forging and Die ··· 27
　4.1　Introduction ·· 27
　4.2　Open-die Forging ··· 27
　4.3　Impression-die and Closed-die Forging ·· 28
　4.4　Precision Forging ··· 29
　4.5　Die Manufacturing Methods ·· 30
　Notes ··· 31
　Glossary ·· 32

Unit 5	**Conventional Machining Processes**	33
5.1	Introduction	33
5.2	Turning and Lathe	33
5.3	Milling and Milling Machine	36
5.4	Drilling and Drill Press	39
	Notes	41
	Glossary	42
Unit 6	**Nontraditional Machining Processes**	45
6.1	Introduction	45
6.2	Electrical Discharge Machining (EDM)	46
6.3	Chemical Machining (CM)	46
6.4	Electrochemical Machining (ECM)	48
6.5	Laser Beam Machining (LBM)	49
6.6	Ultrasonic Machining (USM)	50
	Notes	51
	Glossary	53
Unit 7	**High Speed Cutting (HSC)**	57
7.1	Definition	57
7.2	Introduction to High Speed Cutting	57
7.3	High Speed Cutting Techniques	59
7.4	HSC Machines	59
	Notes	60
	Glossary	61
Unit 8	**Tolerances and Fits**	63
8.1	Introduction	63
8.2	Tolerances	64
8.3	Fits	65
8.4	ISO System of Limits and Fits	66
	Notes	68
	Glossary	69
Unit 9	**Location and Fixtures**	72
9.1	Introduction	72
9.2	Advantages of Jigs and Fixtures	72
9.3	Location of Workpiece	73

    9.4   Clamping of Workpiece ·· 75
    9.5   Classes of Fixtures ·· 76
    Notes ·· 78
    Glossary ·· 79

Unit 10   Gear Transmission ·· 82

    10.1   Introduction ··· 82
    10.2   Spur Gears ·· 82
    10.3   Helical Gears ·· 85
    10.4   Bevel Gears ·· 86
    10.5   Worm Gearing ··· 86
    10.6   Gear Geometry ·· 87
    Notes ·· 90
    Glossary ·· 90

Unit 11   Direct Drive Technology ·· 92

    11.1   Introduction ··· 92
    11.2   Direct-drive Linear (DDL) Motion ·· 92
    11.3   Direct-drive Rotary (DDR) Motors Streamline Machine Design ······················ 94
    11.4   Motorized Spindle ·· 96
    Notes ·· 97
    Glossary ·· 99

Unit 12   Numerical Control ·· 101

    12.1   Introduction ··· 101
    12.2   NC and CNC ·· 101
    12.3   Construction of CNC Machines ·· 102
    12.4   DNC (Distributed Numerical Control) System ·· 104
    Notes ·· 107
    Glossary ·· 108

Unit 13   CNC Machining Centers ··· 109

    13.1   Introduction ··· 109
    13.2   The Concept of Machining Centers ·· 110
    13.3   Types of Machining Centers ··· 110
    13.4   Components of a Machining Center ··· 112
    13.5   Characteristics and Capabilities of Machining Centers ································ 115
    Notes ·· 115

	Glossary	116
Unit 14	**Automatic Control**	**119**
14.1	Introduction	119
14.2	Open-loop Control and Closed-loop Control	119
14.3	Applications of Automatic Control	120
14.4	Artificial Intelligence in Mechatronics	122
	Notes	124
	Glossary	126
Unit 15	**CAD/CAM**	**129**
15.1	Introduction	129
15.2	Geometric Modeling	129
15.3	CAD/CAM	131
15.4	Computer-aided Process Planning	134
	Notes	134
	Glossary	136
Unit 16	**Transducers**	**137**
16.1	Introduction	137
16.2	Transducer Elements	137
16.3	Analog and Digital Transducers	137
16.4	Use of Sensors in Programmable Automation	139
16.5	Some Terms	140
	Notes	141
	Glossary	143
Unit 17	**Robots**	**145**
17.1	Introduction	145
17.2	Definition of Robot	145
17.3	Components of a Robot System	146
17.4	Industrial Robots	147
17.5	Medical Robots	148
17.6	Underwater Robots	149
17.7	Walking Robots	149
17.8	Humanoid Robots	150
17.9	Intelligent Robots	151
	Notes	153

    Glossary ··· 155

Unit 18   Additive Manufacturing ··· 157
    18.1   Introduction ·· 157
    18.2   AM Processes and Materials ·· 158
    18.3   Applications of Additive Manufacturing ··· 161
    18.4   Conclusions ·· 163
    Notes ··· 164
    Glossary ·· 165

Unit 19   New Energy ·· 168
    19.1   Introduction ·· 168
    19.2   Classification of Energy Resources ·· 168
    19.3   New Energy ··· 168
    19.4   Solar Energy ··· 169
    19.5   Nuclear Energy ·· 169
    19.6   Ocean Energy ··· 170
    19.7   Wind Energy ··· 170
    19.8   Biomass Energy (Aka Bioenergy) ·· 171
    19.9   Geothermal Energy ·· 171
    19.10   Hydrogen Energy ·· 171
    19.11   Development Direction of New Energy Vehicles ····································· 172
    Notes ··· 174
    Glossary ·· 175

Unit 20   Development of Modern Manufacturing ··· 178
    20.1   Introduction ·· 178
    20.2   Mechanization ·· 179
    20.3   Programmable Automation ··· 179
    20.4   Computer-aided Manufacturing ··· 180
    20.5   Flexibility ··· 180
    20.6   Remanufacturing ··· 181
    Notes ··· 183
    Glossary ·· 184

Unit 21   IT and Manufacturing ··· 186
    21.1   Introduction ·· 186
    21.2   Computer-integrated Manufacturing ·· 186

21.3 Enterprise Resource Planning (ERP) System ... 188
21.4 Computer-aided System (CAx) ... 189
Notes ... 190
Glossary ... 192

Unit 22 Manufacturing and Roles of Engineers ... 193
22.1 Manufacturing ... 193
22.2 Roles of Engineers ... 195
Notes ... 196
Glossary ... 198

附录 A 科技论文英文摘要写作要点 ... 200
A.1 论文标题 ... 200
A.2 摘要写作注意事项 ... 200
A.3 典型常用语句实例 ... 204
A.4 汉译英范文（参考） ... 205

附录 B 麻省理工学院简介 ... 207
B.1 原文 ... 207
B.2 参考译文 ... 211

附录 C 机床说明书翻译样本 ... 215

附录 D 校园常用词汇 ... 219

附录 E 部分参考译文 ... 223
第 2 单元 工程材料(Ⅱ) ... 223
第 3 单元 铸造 ... 227
第 4 单元 锻造与模具 ... 229
第 6 单元 特种加工工艺 ... 232
第 20 单元 现代制造技术的发展 ... 235
第 21 单元 信息技术与制造业 ... 239

参考文献 ... 242

Unit 1　Engineering Materials (I)

1.1 Introduction

For industrial purposes, materials can be divided into engineering materials and non-engineering materials. Engineering materials are those used in manufacturing and will become parts of products through definite processing. In generally, engineering materials may be further subdivided into metals, ceramics, composites and polymers.

Metals. Common metals are gold, silver, copper, iron, nickel, aluminum, magnesium and titanium, etc. Among these metals, gold and silver (also platinum) are precious metals; iron and nickel are magnetic materials (they are subject to the action of magnetic force); aluminum, magnesium and titanium are commonly called light metals. Of course, metal materials are seldom used in their pure states but in alloy states. Alloys contain more than one metallic element. Their properties can be modified by changing the element contents present in them. Examples of alloys include stainless steels which are alloys of Fe, Ni and Cr; and brass which is an alloy of Cu and Zn. In addition, metal materials can also be broadly divided into two groups, i.e. ferrous metals and nonferrous metals. Ferrous metals often refer to iron alloys (iron and steel materials) and nonferrous metals include all other metallic materials.

Ceramics. It seems that there hasn't been an exact and complete definition about advanced ceramics so far, but from a viewpoint of modern engineering and technology, advanced ceramics (differentiating from traditional household ceramics) may be defined as the new engineering materials composed of some special kinds of metallic oxides (e.g. alumina or corundum and zirconia) or carbides or nitrides of metallic and nonmetallic elements (e.g. tungsten carbide, silicon carbide, boron nitride, silicon nitride, etc.).[1] They have some unique properties such as high-temperature strength; hardness; inertness to chemicals, food, and environment; resistance to wear and corrosion; and low electrical and thermal conductivity. So they are widely used in turbine, automobile, aerospace components, heat exchangers, semiconductors and cutting tools.

Polymers. The word *polymer* was first used in 1866. In essence, they are organic macromolecular compounds. And in 1909, the word *plastics* was employed as a synonym for "polymers". Plastic is from a Greek word *plastikos*, meaning "able to be molded and shaped". Plastics are one of numerous polymeric materials and have extremely large molecules. Because of their many unique and diverse properties, polymers have increasingly replaced metallic components in applications such as automobiles, civilian and military air craft, sporting goods, toys, appliances, and office equipment.

Composite materials. Among the major developments in materials in recent years are composite materials. In fact, composites are now one of the most important classes of engineered materials, because they offer several outstanding properties as compared with conventional materials. A composite material is a combination of two or more chemically distinct and insoluble phases; its properties and structural performance are superior to those of the constituents acting independently.[2]

Nanomaterials refer to those materials, at least one of whose three dimensions is at the nano-scale (1—100 nm) and hence we may have nano-powders, nano-fibers and nano-films.[3] They were first investigated in the early 1980s. However, we can not classify them in nature as distinct from other four common engineering materials, i.e. metals, ceramics, composite and macromolecular materials, because various nano-materials developed so far are all composed of one or combination of the above four materials.[4] Nevertheless, when the sizes of some common materials are reduced to the nano-scale, they will possess some special properties superior to traditional and commercially available materials. These properties can include strength, hardness, ductility, wear resistance and corrosion resistance suitable for structural (load-bearing) and nonstructural applications, in combination with unique electrical, magnetic, and optical properties. For example, nano-powders have very large specific surface area, up to hundreds even thousands of square meters per gram, making them become highly active adsorbents and catalysts with wide application prospect in organic synthesis and environmental protection.[5]

1.2 Ferrous Metals and Alloys

By virtue of their wide range of mechanical, physical, and chemical properties, ferrous metals and alloys are among the most useful of all metals. Ferrous metals and alloys contain iron as their base metal; the general categories are cast irons, carbon and alloy steels, stainless steels, tool and die steels.[6]

The term cast iron refers to a family of ferrous alloys composed of iron, carbon (ranging from 2.11% to about 4.5%), and silicon (up to about 3.5%). Cast irons are usually classified as follows:

1. Gray cast iron, or gray iron;
2. Ductile cast iron, nodular cast iron, or spherical graphite cast iron;
3. White cast iron;
4. Malleable iron;
5. Compacted graphite iron.

The equilibrium phase diagram relevant to cast irons is shown in Fig.1.1, in which the right boundary is 100% carbon, that is, pure graphite. The eutectic temperature is 1154 °C (2109 °F), and so cast irons are completely liquid at temperatures lower than those required for liquid steels. Consequently, iron with high carbon content can be cast at lower temperatures than can steels.

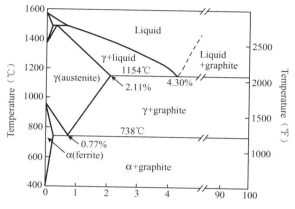

Fig.1.1   Composition graphite(%)

Carbon steels are generally classified by their proportion (by weight) of carbon content.
1. Low-carbon steel, also called mild steel, has less than 0.30% carbon. It is generally used for common industrial products, such as bolts, nuts, sheet, plate, and tubes, and for machine components that do not require high strength.
2. Medium-carbon steel has 0.30% to 0.60% carbon. It is generally used in applications requiring higher strength than is available in low-carbon steels, such as in machinery, in automotive and agricultural equipment parts (gears, axles, connecting rods, crankshafts), in railroad equipment, and in parts for metalworking machinery.
3. High-carbon steel has more than 0.60% carbon. It is generally used for parts requiring strength, hardness, and wear resistance, such as cutting tools, cable, music string, springs, and cutlery. After being manufactured into shapes, the parts are usually heat treated and tempered. The higher the carbon content of the steel, the higher is its hardness, strength, and wear resistance after heat treatment.

Alloy steels contain significant amounts of alloying elements. Structural-grade alloy steels, as identified by ASTM specifications, are used mainly in the construction and transportation industries, because of their high strength. Other alloy steels are used in applications where strength, hardness, creep and fatigue resistance, and toughness are required. These steels may also have been heat treated, in order to obtain the desired properties.

Stainless steels are characterized primarily by their corrosion resistance, high strength and ductility, and high chromium content. They are called *stainless* because in the presence of oxygen (air) they develop a thin, hard adherent film of chromium oxide that protects the metal from corrosion (*passivation*). This protective film builds up again in the event that the surface is scratched.[7] For passivation to occur, the minimum chromium content should be 10% to 12% by weight. In addition to chromium, other alloying elements in stainless steels typically are nickel, molybdenum, copper, titanium, silicon, manganese, columbium, aluminum, nitrogen, and sulfur.

Tool and die steels are specially alloyed steels. They are designed for high strength, impact toughness, and wear resistance at room and elevated temperatures. They are commonly used in forming and machining of metals.

High-speed steels (HSS) are the most highly alloyed tool and die steels. First developed in the early 1900s, they maintain their hardness and strength at elevated operating temperatures. There are two basic types of high-speed steels: the *molybdenum type* (M series) and the *tungsten type* (T series).

Hot-work steels are designed for use at elevated temperatures. They have high toughness as well as high resistance to wear and cracking. The alloying elements are generally tungsten, molybdenum, chromium, and vanadium.

Cold-work steels are used for cold-working operations. They generally have high resistance to wear and cracking. These steels are available as oil-hardening or air-hardening types.

Shock-resisting steels are designed for impact toughness and are used in applications such as header dies, punches, and chisels. Other properties of these steels depend on the particular composition.

1.3 Nonferrous Metals and Alloys

Nonferrous metals and alloys cover a wide range of materials, from common metals such as aluminum, copper, and magnesium to high-strength, high-temperature alloys, such as those of tungsten, tantalum, and molybdenum. Although generally more expensive than ferrous metals, nonferrous metals and alloys have important applications because of properties such as corrosion resistance, high thermal and electrical conductivity, low density, and ease of fabrication.

Aluminum (Al). Typical examples of the applications of nonferrous metals and alloys are aluminum for cooking utensils and aircraft bodies, copper wire for electricity, copper tubing for residential water supply, zinc for galvanized sheet metal for car bodies, titanium for jet-engine turbine blades and for orthopedic implants.

The principal uses of aluminum and its alloys are in containers and packaging (aluminum cans and foil), in buildings and other types of construction, in transportation (aircraft and aerospace applications, buses, automobiles), in electrical applications (economical and nonmagnetic electrical conductor), in consumer durables (appliances, cooking utensils, and furniture). Nearly all high-voltage transmission wiring is made of aluminum. In its structural (load-bearing) components, 82% of a Boeing 747 aircraft (and 79% of a Boeing 757 aircraft) is aluminum.

Porous aluminum blocks of aluminum have recently been produced that are 37% lighter than solid aluminum and have uniform permeability (microporosity). This characteristic allows their use in applications where a vacuum or differential pressure has to be maintained. Examples are the vacuum holding of fixtures for assembly and automation and the vacuum forming or thermoforming of plastics.[8] These blocks are 70% to 90% aluminum powder; the rest is epoxy resin. They can be machined with relative ease and can be joined together using adhesives.

Magnesium (Mg) is the lightest engineering metal available, and it has good vibration-damping characteristics. Its alloys are used in structural and nonstructural applications wherever weight is of primary importance. Magnesium is also an alloying element in various nonferrous metals. A variety of magnesium alloys have good casting, forming, and machining characteristics.

Typical uses of magnesium alloys are in aircraft and missile components, material-handling equipment, portable power tools (such as drills and sanders), ladders, luggage, bicycles, sporting goods, and general lightweight components.

Copper (Cu), first produced in about 4000 B.C., and its alloys have properties somewhat similar to those of aluminum and its alloys. In addition, they are among the best conductors of electricity and heat, and they have good corrosion resistance. They can be processed easily by various forming, machining, casting, and joining techniques.

Copper alloys are often attractive for applications where a combination of electrical, mechanical, nonmagnetic, corrosion-resistant, thermally conductive, and wear-resistant qualities are required. Applications include electrical and electronic components; springs; cartridges for small arms; plumbing; heat exchangers; marine hardware, and consumer goods, such as cooking utensils, jewelry, and other decorative objects.

Titanium (Ti), named after the giant Greek god Titan, was discovered in 1791, but it was not produced commercially until the 1950s. Although it is expensive, its high strength-to-weight ratio and its corrosion resistance at room and elevated temperatures make it attractive for many applications including aircraft, jet-engine, racing-car, chemical, petrochemical, and marine components, submarine hulls, and biomaterials, such as orthopedic implants. Titanium alloys have been developed for service at 550 °C (1000 °F) for long periods of time and at up to 750 °C (1400 °F) for shorter periods.

The properties and manufacturing characteristics of titanium alloys are extremely sensitive to small variations in both alloying and residual elements. These elements cause embrittlement of titanium and, consequently, reduce toughness and ductility.

Superalloys are important in high-temperature applications; hence, they are also known as heat-resistant or high-temperature alloys. Major applications of superalloys are in jet engines and gas turbines; other applications are in reciprocating engines, in rocket engines, in tools and dies for hot-working of metals, and in the nuclear, chemical, and petrochemical industries. Superalloys generally have good resistance to corrosion, to mechanical and thermal fatigue, to mechanical and thermal shock, to creep, and to erosion at elevated temperatures.

Most superalloys have a maximum service temperature of about 1000 °C (1800 °F) in structural applications. The temperatures can be as high as 1200 °C (2200 °F) and above and a major application for the superalloys of rapidly-solidified powders is consolidation into near-net shapes for parts used in aerospace engines.

Low-melting alloys are so named because of their relatively low melting points. The major metals in this category are lead, zinc, and tin and their alloys.

Lead (Pb) has properties of high density, resistance to corrosion (by virtue of the stable lead-oxide layer that forms to protect the surface), softness, low strength, ductility, and good workability. Lead is also used for damping sound and vibrations, in radiation shielding against X-rays, in ammunition, as weights, and in the chemical industry. An additional use of lead is as a solid lubricant for hot-metal forming operations. Because of its toxicity, however, environmental contamination by lead (causing lead poisoning) is a major concern.

Zinc (Zn), bluish-white in color, is the metal fourth most utilized industrially, coming after iron, aluminum, and copper. Zinc is also used as an alloying element. Brass, for example, is an alloy of copper and zinc. Major alloying elements in zinc-based alloys are aluminum, copper, and magnesium. Zinc-based alloys are used extensively in die casting, for making such products as fuel pumps and grills for automobiles, components for household appliances such as vacuum cleaners, washing machines, and kitchen equipment, and various machinery parts and photoengraving equipment.

Although used in small amounts, *tin* (Sn) is an important metal. The most extensive use of tin, a silvery-white, lustrous metal, is as a protective coating on the steel sheet (tin plate) that is used in making containers (tin cans) for food and for various other products. Unalloyed tin is used in such applications as lining material for water distillation plants and as a molten layer of metal over which plate glass (float glass) is made. Tin-based alloys (also called white metals) generally contain copper, antimony, and lead. The alloying elements impart hardness, strength, and corrosion resistance.

Precious metals. Gold, silver, and platinum are the most important precious (that is, costly) metals; they are also called noble metals.

    1. *Gold* (Au) is soft and ductile, and it has good corrosion resistance at any temperature. Typical applications include jewelry, coinage, reflectors, dental work, electroplating,

and electrical contacts and terminals.
2. *Silver* (Ag) is a ductile metal, and it has the highest electrical and thermal conductivity of any metal. It does, however, develop an oxide film that affects its surface characteristics and appearance. Typical applications for silver include tableware, jewelry, coinage, electroplating, photographic film, electrical contacts, solders, bearing linings, and food and chemical equipment.
3. *Platinum* (Pt) is a soft, ductile, grayish-white metal that has good corrosion resistance even at elevated temperatures. Platinum alloys are used as electrical contacts, for spark-plug electrodes, as catalysts for automobile pollution-control devices, in filaments, in nozzles, in dies for extruding glass fibers, in thermocouples, in the electrochemical industry, as jewelry, and in dental work.

Notes

[1] It seems that there hasn't been a exact and complete definition about advanced ceramics so far, but from a viewpoint of modern engineering and technology, advanced ceramics (differentiating from traditional household ceramics) may be defined as the new engineering materials composed of some special kinds of metallic oxides (e.g. alumina or corundum and zirconia) or carbides or nitrides of metallic and nonmetallic elements (e.g. tungsten carbide, silicon carbide, boron nitride, silicon nitride etc.).

句意：迄今为止，似乎还没有一个关于先进陶瓷材料的确切而完整的定义。但按照现代工程技术的观点，与传统的家用陶瓷材料不同的是，我们可以将先进陶瓷材料定义为由某些特殊金属氧化物［如氧化铝（或称为刚玉）和氧化锆］，或者金属和非金属元素的碳化物或氮化物（如碳化钨、碳化硅、氮化硼、氮化硅等）组成的一类新型工程材料。

[2] A composite material is a combination of two or more chemically distinct and insoluble phases; its properties and structural performance are superior to those of the constituents acting independently.

句意：复合材料是由两种或两种以上化学性质完全不同，且能相溶的组分所构成的，它的性质和结构特征要比各组成相单独作用时更加优越。

[3] Nanomaterials refer to those materials, at least one of whose three dimensions is at the nano-scale (1—100nm) and hence we may have nano-powders, nano-fibers and nano-films.

句意：纳米材料指的是这样一类材料：在它们的三维尺寸中至少有一维是在纳米尺度（1～100 nm）以内，因而有所谓的纳米粉、纳米纤维和纳米薄膜。

[4] However, we can not classify them in nature as distinct from other four common engineering materials, i.e. metals, ceramics, composite and macromolecular materials because various nano-materials developed so far are all composed of one or combination of the above four materials.

句意：然而，我们却不能将它们在本质上与其他4种常用的工程材料即金属、陶瓷、复合材料和高分子材料区分开来，因为迄今为止所开发的各种纳米材料都不外乎是由以上4种工程材料中的一种或几种组合而成的。

[5] For example, nano-powders have very large specific surface area, up to hundreds even thousands of square meters per gram, making them become highly active adsorbents and catalysts with wide application prospect in organic synthesis and environmental protection.

句意：例如，纳米粉具有极大的比表面积，甚至可达每克数百甚至数千平方米，使它们成为活性极高的吸附剂或催化剂，在有机合成和环境保护领域具有广阔的应用前景。

[6] ferrous metal 和 nonferrous metal 按英文原意是铁类（基）金属和非铁类（基）金属。现在工程界将其译成黑色金属和有色金属是错误的！英美工程界从来不承认金属分为黑色和有色，只认为金属分为铁基金属和非铁基金属。所以除钢铁材料外，所有其他一切金属统统是 nonferrous 金属！之所以发生这样的错译是20世纪50年代照搬苏联教科书的结果。当时俄语中将铁和常与铁形成共生矿的锰及铬归类为黑色金属 чёрный металл，而将其他金属归类为有色金属 цветной металл。所以我国的工程界也将铁、锰、铬作为黑色金属，其他金属则归类为有色金属。因此将英文的 ferrous 译成黑色，nonferrous 译成有色。现在如果查百度上的关键词"黑色金属"，就是如此解释的，它包括铁、锰和铬。

[7] Stainless steels are characterized primarily by their corrosion resistance, high strength and ductility, and high chromium content. They are called *stainless* because in the presence of oxygen (air) they develop a thin, hard adherent film of chromium oxide that protects the metal from corrosion (*passivation*). This protective film builds up again in the event that the surface is scratched.

句意：不锈钢的主要特点是它们具有很好的耐蚀性，强度高，延展性好，铬含量高。之所以被称为不锈钢，是因为在有氧环境下会生成一层牢牢附着在表面的氧化铬薄膜，从而保护其不被腐蚀（所谓钝化作用）。即使金属表面受到刮擦而破损，这层保护膜还会继续生成。

[8] Porous aluminum blocks of aluminum have recently been produced that are 37% lighter than solid aluminum and have uniform permeability (microporosity). This characteristic allows their use in applications where a vacuum or differential pressure has to be maintained. Examples are the vacuum holding of fixtures for assembly and automation and the vacuum forming or thermoforming of plastics.

句意：多孔铝是最近才生产出的新材料，因为含有均匀分布的细小微孔，所以它比实体铝块轻37%。这种特性使它们可在需要保持真空或压差的环境下得到应用（由于可通过微孔非常均匀和方便地抽成真空）。例如，用于装配和自动化操作的真空吸紧夹具，以及塑料的真空成形和热成形工艺。

Glossary

alumina   *n.* 氧化铝，刚玉

cartridge   *n.* 夹头，支架，支座

catalyst   *n.* 催化剂

compacted graphite iron   蠕墨铸铁

composite   *n.* 复合材料

embrittlement   *n.* 脆化

ferrous metal   铁类（基）金属（我国工程界译为黑色金属）

galvanizing   *n.* 镀锌

hardenability   *n.* 可淬性

high-speed steel (HSS)   高速钢

luster   *n.* 光泽

macromolecular compound   大分子化合物

malleable iron   可锻铸铁

nanomaterial   *n.* 纳米材料

nonferrous metal   非铁类（基）金属（我国工程界译为有色金属）

passivation   *n.* 钝化

polymer   *n.* 聚合物

porous aluminum   多孔铝材

precious metal (noble metal)   贵金属

solder   *n.* 焊料

specific surface   比表面积

superalloy   *n.* 超级合金，耐热合金

zirconia   *n.* 氧化锆

Unit 2 Engineering Materials (II)

2.1 Introduction

In unit one, we briefly introduced some metallic materials. In this section, we will give an outline of three other commonly used engineering materials, i.e. ceramics, polymers and composites.

2.2 Ceramics

Ceramics are compounds of metallic and nonmetallic elements. Because of the large number of possible combinations of elements, a great variety of ceramics are now available for a wide range of consumer and industrial applications. Ceramics have been used for many years in automotive spark plugs, as an electrical insulator and for high-temperature strength. They have become increasingly important in tool and die materials, in heat engines, and in automotive components such as exhaust-port liners, coated pistons, and cylinder liners.

Ceramics can be divided into two general categories:
1. Traditional, such as whiteware, tiles, brick, sewer pipe, pottery, and abrasive wheels.
2. Industrial ceramics, also called engineering, fine ceramics, or advanced ceramics, have found wide applications in turbine, automobile, aerospace components, heat exchangers, semiconductors, seals, prosthetics and cutting tools.

Ceramics are available as a *single crystal* or in *polycrystalline* form, consisting of many grains. Grain size has a major influence on the strength and properties of ceramics, the finer the grain size, the higher are the strength and toughness.

Oxide Ceramics. There are two major types of oxide ceramics: alumina and zirconia.

Alumina. Also called corundum or emery, *alumina* (aluminum oxide, Al_2O_3) is the most widely used *oxide ceramic*, either in pure form or as a raw material to be mixed with other oxides. It has high hardness and moderate strength. Although alumina exists in nature, it contains unknown amounts of impurities and possesses non-uniform properties. As a result, its behavior is unreliable. Aluminum oxide, silicon carbide, and many other ceramics are now manufactured almost totally synthetically, so that their quality can be controlled.

Zirconia. Zirconia (zirconium oxide, ZrO_2, white in color) has good toughness; good resistance to thermal shock, wear, and corrosion; low thermal conductivity; and a low friction coefficient. Partially stabilized zirconia (PSZ) has high strength and toughness and better reliability in performance than does zirconia. It is obtained by doping the zirconia with oxides of calcium, yttrium, or magnesium.[1] Typical applications include dies for hot extrusion of metals and zirconia beads used as grinding and dispersion media for aerospace coatings, for automotive primers and topcoats, and for fine glossy print on flexible food packaging.

Other ceramics may be classified as follows:

Carbides. Typical examples of carbides are those of tungsten (WC) and titanium (TiC), used as cutting tools and die materials, and that of silicon (SiC), used as an abrasive (especially in grinding wheels).

1. *Tungsten carbide* consists of tungsten-carbide particles with cobalt as a binder. The amount of binder has a major influence on the material's properties. Toughness increases with cobalt content, whereas hardness, strength, and wear resistance decrease.
2. *Titanium carbide* has nickel and molybdenum as the binder and is not as tough as tungsten carbides.
3. *Silicon carbide* has good resistance to wear, thermal shock, and corrosion. It has a low friction coefficient, and it retains strength at elevated temperatures. It is suitable for high-temperature components in heat engines and is also used as an abrasive.

Nitrides. Another important class of ceramics is the nitrides, particularly cubic boron nitride (CBN), titanium nitride (TiN), and silicon nitride (Si_3N_4).

1. *Cubic boron nitride*, the second hardest known substance (after diamond), has special applications, such as in cutting tools and for abrasives in grinding wheels. It does not exist in nature; it was first made synthetically in the 1970s, by means of techniques similar to those used in making synthetic diamond.
2. *Titanium nitride* is used widely as a coating on cutting tools. It improves tool life by virtue of its low frictional characteristics.
3. *Silicon nitride* has high resistance to creep at elevated temperatures, low thermal expansion, and high thermal conductivity; consequently, it resists thermal shock. It is suitable for high-temperature structural applications, such as in automotive engine and gas-turbine components, in cam-follower rollers, in bearings, in sand-blast nozzles, and in components for the paper industry.

Cermets. Cermets are combinations: a *ceramic* phase bonded with a *metallic* phase. Introduced in the 1960s and also called black ceramics or hot-pressed ceramics, they combine the high-temperature oxidation resistance of ceramics with the toughness, thermal-shock resistance, and ductility of metals.[2] An application of cermets is in cutting tools, a typical composition being 70% aluminum oxide and 30% titanium carbide.

Other cermets contain various oxides, carbides, and nitrides. They have been developed for high-temperature applications, such as nozzles for jet engines and brakes for aircraft. Cermets can be regarded as composite materials; they can be used in various combinations of ceramics and metals bonded by powder-metallurgy techniques.

The capability of ceramics to maintain their strength and stiffness at elevated temperatures makes them very attractive for high-temperature applications. Their high resistance to wear makes them suitable for applications such as cylinder liners, bushings, seals, and bearings. The higher operating temperatures made possible by the use of ceramic components mean more efficient burning of fuel and reduction of emissions in automobiles. Currently, internal combustion engines are only about 30% efficient, but with the use of ceramic components the operating performance can be improved by at least 30%.

Much research has been conducted on developing materials and techniques for an all-ceramic heat engine capable of operating at temperatures up to 1000 °C (1830 °F). The development of such an engine has, however, been slower than expected because of such problems as unreliability, lack of sufficient toughness, difficulty with lubricating bearings and hot components, and a lack of the capability for structural ceramics (such as silicon nitride and silicon carbide) to be produced economically in near-net shape, as weighed against the need for the machining and finishing processes required for dimensional accuracy of the engine.[3]

Bioceramics. Because of their strength and inertness, ceramics are used as biomaterials to replace joints in the human body, as prosthetic devices, and in dental work. Furthermore, ceramic implants can be made porous; bone can grow into the porous structure (likewise with porous titanium implants) and develop a strong bond, having high structural integrity between them. Commonly used bioceramics are aluminum oxide, silicon nitride, and various compounds of silica.

2.3 Polymers

Polymeric materials have extremely large molecules (*macromolecules*). Consumer and industrial products made of polymers include food and beverage containers, packaging, signs, textiles, medical devices, foams, paints, safety shields, toys, appliances, lenses, gears, electronic and electrical products, and automobile bodies and components. They have many unique and diverse properties:

    1. corrosion resistance and resistance to chemicals;

    2. low electrical and thermal conductivity;

    3. low density;

    4. high strength-to-weight ratio, particularly when reinforced;

    5. noise reduction;

    6. wide choice of colors and transparencies.

The development of modern plastics technology began in the 1920s, when the raw materials necessary for making polymers were extracted from coal and petroleum products. Ethylene was the first example of such a raw material; it became the building block for polyethylene. Ethylene is the product of the reaction between acetylene and hydrogen, and acetylene is the product of the reaction between coke and methane. The commercial polymers, including polypropylene, polyvinyl chloride, polycarbonate, and others, are all made in a similar manner; these materials are known as synthetic organic polymers.

Thermoplastics. When we heat the polymer and then cool the polymer, it returns to its original hardness and strength; in other words, the process is reversible, the polymer is known as thermoplastics, typical examples of which are acrylics, cellulosics, nylons, polyethylenes, and polyvinyl chloride.

Thermosets. However, if the long-chain molecules in a polymer are cross-linked in a three-dimensional arrangement, the structure in effect becomes one *giant molecule* with strong covalent bonds, these polymers are called thermosetting polymers, or thermosets, because, during polymerization, the network is completed and the shape of the part is permanently set.[4] This curing (cross-linking) reaction, unlike that of thermoplastics, is *irreversible*. If the temperature is increased sufficiently, the thermosetting polymer begins to burn up, degrade, and char. Thermosetting plastics generally possess better mechanical, thermal, and chemical properties, electrical resistance, and dimensional stability than do thermoplastics. A typical thermoset is *phenolic*, which is a product of the reaction between phenol and formaldehyde. Common products made from this polymer are the handles and knobs on cooking pots and pans and components of light switches and outlets.

Biodegradable plastics. Plastic wastes contribute about 10% of municipal solid waste. One-third of plastic production goes into disposable products, such as bottles, packaging, and garbage bags. With the growing use of plastics, and with increasing concern over environmental issues regarding the disposal of plastic products and the shortage of landfills, major efforts are underway to develop completely biodegradable plastics. Most plastic products have traditionally been made from synthetic polymers that are derived from nonrenewable natural resources, are not biodegradable, and are difficult to recycle. Biodegradability means that microbial species in the environment (e.g. microorganisms in soil and water) will degrade a portion of (or even the entire) polymeric material, under the right environmental conditions, and without producing toxic by-products.[5] The end products of the degradation of the biodegradable portion of the material are carbon dioxide and water. Because of the variety of constituents in biodegradable plastics, these plastics can be regarded as composite materials; consequently, only a portion of these plastics may be truly biodegradable.

2.4　Composite Materials

Among the major developments in materials in recent years are composite materials. In fact, composites are now one of the most important classes of engineered materials, because they offer several outstanding properties as compared with conventional materials. A composite material is a combination of two or more chemically distinct and insoluble phases; its properties and structural performance are superior to those of the constituents acting independently.

For example, it is known that plastics possess mechanical properties that are generally inferior to those of metals and alloys—in particular, low strength, stiffness, and creep resistance. These properties can be improved by embedding reinforcements of various types (such as glass or graphite fibers) to produce reinforced plastics. Metals and ceramics, as well, can be embedded with particles or fibers, to improve their properties; these combinations are known as metal-matrix and ceramic-matrix composites.[6]

Applications of reinforced plastics. The first application of reinforced plastics (in 1907) was for an acid-resistant tank, made of a phenolic resin with asbestos fibers. Epoxies were first used as a matrix material in the 1930s. Beginning in the 1940s, boats were made with fiberglass. Major developments in composites began in the 1970s; those materials are now called advanced composites. They are typically used in military and commercial aircraft and in rocket components, helicopter blades, automobile bodies, leaf springs, drive shafts, pipes, ladders, pressure vessels, sporting goods, sports and military helmets, boat hulls, and various other structures and components. Glass- or carbon-fiber reinforced hybrid plastics are now being developed for high-temperature applications, with continuous use at about 300 °C (550 °F). The Boeing 777 is made of about 9% composites by total weight; that proportion is triple the composite content of previous Boeing transport aircraft. The floor beams and panels and most of the vertical and horizontal tail are made of composite materials. By virtue of the resulting weight savings, reinforced plastics have reduced fuel consumption by about 2%.

Metal-matrix composites (MMC). The advantages of a metal matrix over a polymer matrix are its higher elastic modulus, its resistance to elevated temperatures, and its higher toughness and ductility. The limitations are higher density and greater difficulty in processing parts. Matrix materials in these composites are usually aluminum, aluminum–lithium, magnesium, copper, titanium, and superalloys. Fiber materials can be graphite, aluminum oxide, silicon carbide, boron, molybdenum, and tungsten. The elastic modulus of nonmetallic fibers ranges between 200 GPa and 400 GPa, with tensile strengths being in the range from 2000 MPa to 3000 MPa. Because of their high specific stiffness, light weight, and high thermal conductivity, boron fibers in an aluminum matrix have been used for structural tubular supports in the Space Shuttle orbiter.

Ceramic-matrix composites (CMC). Composites with a ceramic matrix are another important development in engineered materials because of their resistance to high temperatures and corrosive environments. Ceramics are strong and stiff, and they resist high temperatures, but they generally lack toughness. Matrix materials that retain their strength up to 1700 °C (3000 °F) are silicon carbide, silicon nitride and aluminum oxide. Various techniques for improving the mechanical properties of ceramic-matrix composites, particularly their toughness, are being investigated. Applications are in jet and automotive engines, deep-sea mining equipment, pressure vessels, structural components, cutting tools, and dies for the extrusion and drawing of metals.

2.5 Metamaterials

Introduction. Metamaterials refer to those materials that possess man-designed structures and unusual even supernormal properties far beyond those of their compositions or other natural or artificial materials.[7] The word "Meta" is taken from Greek whose meaning is "beyond" or "super". Metamaterials have some special properties beyond the naturally occurring materials. These are the materials that extract their properties from their structure rather than the material of which they are composed of. Significant advances in optical, acoustical and thermal metamaterials have inspired the study of mechanical metamaterials. Mechanical metamaterials exhibit extraordinary properties such as negative compressibility, negative Poisson's ratio, negative mass density and elastic modulus and exceptional elastodynamic behaviors.

American Journal of Science lists them as one of the top ten scientific advances in the first 10 years in the century, which may trigger a new round of technological revolution in IT, defense industry, new energy, micro-machining, etc.

Metamaterial types. Generally speaking, metamaterials can be classified into mechanical metamaterials, electromagnetic metamaterials, and acoustical metamaterials, etc. There have been intensive research activities on these materials during the past decade. Since metamaterials significantly enlarge material space available in designing wave-control devices, extensive investigations were carried out in the corresponding resonant mechanism, dynamic homogenization theory and microstructural design. The rapid advances of metamaterials lead to a number of potential applications, perhaps the most fantastic metamaterials are materials especially engineered to have a peculiar physical behavior, to be exploited for some well-specified technological applications. Most attractive examples are the invisible cloak and the super lens.[8] Artificially constructed metamaterials have become useful tools in optics and electromagnetics for the construction of devices with complex spatial- or frequency-domain behaviors. Recent developments in reconfigurable and tunable metamaterials have extended the possibility for fabricating metadevices and unique, subwavelength devices with practical functionality. In addition to exhibiting electromagnetic responses not readily available in nature, these metadevices offer the possibility for improved performance characteristics in smaller, multifunctional applications.

Notes

[1] Partially stabilized zirconia (PSZ) has high strength and toughness and better reliability in performance than does zirconia. It is obtained by doping the zirconia with oxides of calcium, yttrium, or magnesium.

句意：部分稳定氧化锆具有较高的强度和韧性，比氧化锆性能更可靠。它是通过对氧化锆进行钙、钇和镁的氧化物掺杂制成的一种新型陶瓷。

[2] Cermets are combinations: a *ceramic* phase bonded with a *metallic* phase. Introduced in the 1960s and also called black ceramics or hot-pressed ceramics, they combine the high-temperature oxidation resistance of ceramics with the toughness, thermal-shock resistance, and ductility of metals.

句意：金属陶瓷是 20 世纪 60 年代出现的一种由陶瓷相和金属相构成的混合物，又称为黑色陶瓷或热压陶瓷。它们将陶瓷的抗高温氧化性，以及金属材料的韧性、抗热振性和延展性结合在一起。

[3] The development of such an engine has, however, been slower than expected because of such problems as unreliability, lack of sufficient toughness, difficulty with lubricating bearings and hot components, and a lack of the capability for structural ceramics (such as silicon nitride and silicon carbide) to be produced economically in near-net shape, as weighed against the need for the machining and finishing processes required for dimensional accuracy of the engine.

句意：然而，由于性能的不稳定、缺乏足够的韧性、对轴承和热工作部件的润滑困难，并考虑到发动机的尺寸精度、要求最后进行机加工和精整加工，从而缺乏低成本生产近净成形的结构陶瓷（如氮化硅和碳化硅）的能力，因此这种全陶瓷结构的热力发动机的发展进程最终比预期要慢。

[4] However, if the long-chain molecules in a polymer are cross-linked in a three-dimensional arrangement, the structure in effect becomes one giant molecule with strong covalent bonds, these polymers are called thermosetting polymers, or thermosets, because, during polymerization, the network is completed and the shape of the part is permanently set.

句意：然而，如果聚合物的长链分子交联成三维结构，就会成为大分子。这种大分子具有一种强有力的共价键结构，这时的聚合物就成为了热固性聚合物。因为聚合反应时形成了网状结构，所以聚合后的零件形状也就永久保持不变了。

[5] Biodegradability means that microbial species in the environment (e.g., microorganisms in soil a nd water) will degrade a portion of (or even the entire) polymeric material, under the right environmental conditions, and without producing toxic by-products.

句意：聚合物的生物降解性是指环境中的微生物物种（例如，土壤和水中的微生物）在合适的条件下可以降解部分或全部聚合物材料，而不生成有毒的副产品。

[6] These properties can be improved by embedding reinforcements of various types (such as glass or graphite fibers) to produce reinforced plastics. Metals and ceramics, as well, can be embedded with plastics or fibers, to improve their properties; these combinations are known as metal-matrix and ceramic-matrix composites.

句意：这些性能可以通过各种形式的包埋增强（例如，加入玻璃或石墨纤维）来得到改善，从而生产出增强塑料。金属和陶瓷也可以采用加入塑料或纤维的方式来提高其性能。这种材料被称为金属基和陶瓷基复合材料。

[7] Metamaterials refer to those materials that possess man-designed structures and unusual even supernormal properties far beyond those of their compositions or other natural or artificial materials.

句意：超材料指那些由人工设计其结构并具有非同一般甚至极不寻常特性的材料，其性能远远超过了它们的组成成分或其他自然或人造材料。

[8] The rapid advance of metamaterials lead to a number of potential applications, perhaps the most fantastic metamaterials are materials especially engineered to have a peculiar physical behavior, to be exploited for some well-specified technological applications. Most attractive examples are the invisible cloak and the super lens.

句意：超材料的迅速发展使它们发挥出巨大的应用潜力，也许最不可思议的就是采用工程化手段专门制成的超材料，它们具有特殊的物理性质，开发这类材料专门应用于某些特殊的技术领域。其中最吸引人的例子就是隐身斗篷和超级透镜。

Glossary

abrasive wheel　砂轮
acoustical　*adj.* 声学的，听觉的，音响的
amplifying　*adj.* 放大的
artificial　*adj.* 人造的，仿造的，假的
bioceramics　*n.* 生物陶瓷

biodegradable plastics 可降解塑料
ceramic-matrix composite 陶瓷基复合材料
cermet *n.* 金属陶瓷
compressibility *n.* 压缩性，压缩系数，压缩率
cubic boron nitride (CBN) 立方氮化硼
cylinder liner 汽缸套
doping *n.* 掺杂
elastic modulus 弹性模量
elastodynamic 弹性动力学的
electromagnetic *adj.* 电磁的
emery *n.* 刚玉，金刚砂
embedding reinforcement （材料）包埋增强
embedded inclusion 嵌入（包埋的）杂质
ethylene *n.* 乙烯
exhaust-port liner 排气口内衬
fantastic *adj.* 奇异的，古怪的，极好的，不可思议的
harnessing(harness 的现在分词) 治理，利用
heat engine 热力发动机
homogenization *n.* 均质化，均化作用
homogenous *adj.* 同质的，同类的，[数学]齐次的
insulator *n.* 绝缘子
invisible cloak 隐身斗篷
isotropic adj 各向同性的，等方性的
metadevice 超装置，超器件
metamaterial 超材料
matrix *n.* 基体
metal-matrix composite 金属基复合材料
micro/nanoscale 微/纳米尺度的
near-net shape 近净成形
polycarbonate *n.* 聚碳酸酯
polycrystalline *adj.* 多晶的
polypropylene *n.* 聚丙烯
polyvinyl chloride 聚氯乙烯
Poisson's ratio 泊松比
prosthetics *n.* 医疗修复，假肢安装
reconfigurable *adj.* 可重构的，可重组的

reinforced plastics   增强塑料
resonant   *adj.* 共振的，共鸣的
silicon nitride   氮化硅
spark plug   火花塞
subwavelength   子波长，亚波长
super lens   超级透镜
supernormal   *adj.* 非凡的，异于寻常的
thermoplastics   *n.* 热塑性材料
thermoset   *n.* 热固性材料
thermosetting polymer   热固性聚合物
titanium carbide   碳化钛
trapping   *n.* 俘获；设陷阱，使密封于管内
tunable   *adj.* 可调谐的，可调音的
tungsten carbide   碳化钨
whiteware   *n.* 白色陶瓷，卫生陶瓷
zirconia   *n.* 氧化锆

Unit 3　Casting

3.1　Introduction

Several different methods, such as casting, forging, welding and machining are available to shape metals into useful products. Customarily, casting, forging, welding, etc., are referred to as forming processes which generally use molds, dies or other tooling to force the molten metal or metal blanks shaped into final parts. Machining (including turning, milling, shaping, grinding and drilling, etc.), however, means removing the unwanted material from the blank surface in the form of chips with cutting tools or other physical and chemical means.[1] One of the oldest forming processes is casting, which basically involves pouring molten metal into a mold cavity where, upon solidification, it takes the shape of the cavity. Casting was first used around 4000 B.C. to make ornaments, copper arrowheads, and various other objects. The casting process is capable of producing intricate shapes in one piece, including those with internal cavities, such as engine blocks. Almost all metals can be cast to the final shape desired, often with only minor finishing operations required. This capability places casting among the most important net-shape manufacturing technologies.

3.2　Sand Casting

The traditional method of casting metals is in sand molds and has been used for millennia. Simply stated, *sand casting* consists of (1) placing a pattern having the shape of the desired casting in sand to make an imprint, (2) incorporating a gating system, (3) filling the resulting cavity with molten metal, (4) allowing the metal to cool until it solidifies, (5) breaking away the sand mold, and (6) removing the casting.

The following are the major components of sand molds:
1. The mold itself, which is supported by a flask. Two-piece molds consist of a cope on top and a drag on the bottom. The seam between them is the parting line. When more than two pieces are used, the additional parts are called *cheeks*.
2. A pouring basin or *pouring cup*, into which the molten metal is poured.
3. A sprue, through which the molten metal flows downward.
4. The runner system, which has channels that carry the molten metal from the sprue to the mold cavity. Gates are the inlets into the mold cavity.
5. Risers, which supply additional metal to the casting as it shrinks during solidification.

Fig.3.1 shows two different types of risers: a *blind riser* and an *open riser*.

6. Cores, which are inserts made from sand. They are placed in the mold to form hollow regions or otherwise define the interior surface of the casting. Cores are also used on the outside of the casting to form features such as lettering on the surface of a casting or deep external pockets.
7. Vents, which are placed in molds to carry off gases produced when the molten metal comes into contact with the sand in the mold and core. They also exhaust air from the mold cavity as the molten metal flows into the mold.

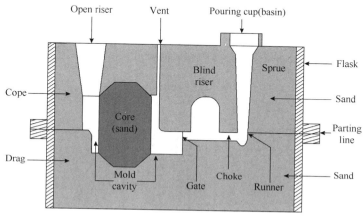

Fig.3.1 Schematic illustration of a sand mold

Patterns are used to mold the sand mixture into the shape of the casting. They may be made of wood, plastic, or metal. The selection of a pattern material depends on the size and shape of the casting, the dimensional accuracy, the quantity of castings required, and the molding process.

3.3 Investment Casting

The investment-casting process, also called the lost-wax process, was first used during the period 4000—3000 B.C. The pattern is made of wax or plastic (such as polystyrene) by molding or rapid prototyping techniques.

The sequences involved in investment casting are shown in Fig.3.2. The pattern is made by injecting molten wax or plastic into a metal die in the shape of the pattern. The pattern is then dipped into a slurry of refractory material such as very fine silica and binders, including water, ethyl silicate, and acids. After this initial coating has dried, the pattern is coated repeatedly to increase its thickness.[2] The term investment derives from the fact that the pattern is invested (covered) with the refractory material. Wax patterns require careful handling because they are not strong enough to withstand the forces involved during mold making. However, unlike plastic patterns (except thermoplastics), wax can be recovered and reused. When the wax pattern is melted away and the mold cavity is left, the molten metal is poured into the cavity, forming the final casting. Intricate shapes can be made with high accuracy. In addition, metals that are hard to machine or fabricate are good candidates for this process.

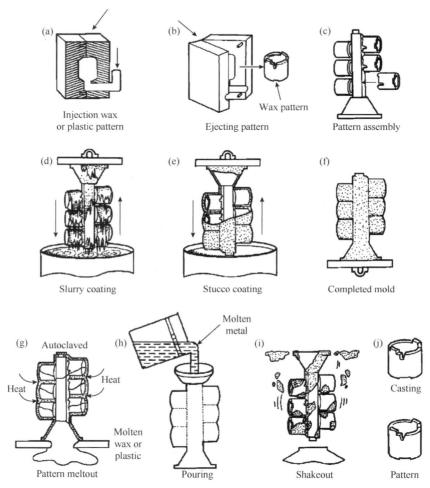

Fig.3.2 Schematic illustration of investment casting

3.4 Expendable-pattern Casting

The expendable-pattern casting process uses a polystyrene pattern, which evaporates upon contact with molten metal to form a cavity for the casting. The process is also known as evaporative-pattern or lost-pattern casting, and under the trade name Full-Mold process.[3] It was formerly known as the "expanded polystyrene process, " and has become one of the more important casting processes for ferrous and nonferrous metals, particularly for the automotive industry.

In this process, raw expendable polystyrene (EPS) beads, containing 5% to 8% pentane (a volatile hydrocarbon), are placed in a preheated die which is usually made of aluminum. The polystyrene expands and takes the shape of the die cavity; additional heat is applied to fuse and bond the beads together. The die is then cooled and opened, and the polystyrene pattern is removed. Complex patterns may also be made by bonding various individual pattern sections using hot-melt adhesive.

The pattern is coated with a water-based refractory slurry, dried, and placed in a flask. The flask is then filled with loose fine sand, which surrounds and supports the pattern (Fig.3.3) and may be dried or mixed with bonding agents to give it additional strength. The sand is compacted around the pattern by various means. Then, without removing the polystyrene pattern, the molten metal is poured into the mold, immediately vaporizes the pattern (an ablation process) and fills the mold cavity, completely replacing the space previously occupied by the polystyrene pattern. The heat degrades (depolymerizes) the polystyrene and the degradation products are vented into the surrounding sand.

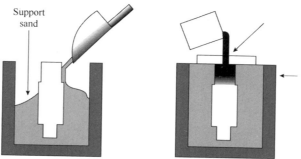

Fig.3.3　Schematic illustration of the expendable-pattern casting process

3.5　Centrifugal Casting

As its name implies, the centrifugal-casting process utilizes the inertial forces caused by rotation to distribute the molten metal into the mold cavities. This method was first suggested in the early 1800s. Hollow cylindrical parts, such as pipes, gun barrels, and street lamp posts, are produced by the technique shown in Fig.3.4, in which molten metal is poured into a rotating mold. The axis of rotation is usually horizontal. Molds are made of steel, iron, or graphite, and may be coated with a refractory lining to increase mold life.

The mold surfaces can be shaped so that pipes with various outer shapes, including square or polygonal, can be cast. The inner surface of the casting remains cylindrical because the molten metal is uniformly distributed by centrifugal forces. However, because of density differences, lighter elements such as dross, impurities, and pieces of the refractory lining tend to collect on the inner surface of the casting. Cylindrical parts ranging from 13 mm (0.5 in.)

to 3 m (10 ft) in diameter and 16 m (50 ft) long can be cast centrifugally, with wall thicknesses ranging from 6 mm to 125 mm (0.25 in. to 5 in.). The pressure generated by the centrifugal force is high, and such high pressure is necessary for casting thick-walled parts. Castings of good quality, dimensional accuracy, and external surface detail are obtained by this process. In addition to pipes, typical parts made are bushings, engine cylinder liners, and bearing rings with or without flanges.

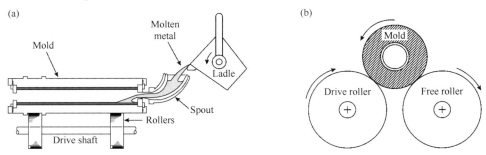

Fig.3.4 Schematic illustration of the centrifugal casting process

3.6 Inspection of Casting

Several methods can be used to inspect castings to determine their quality and the presence of any defects. Castings can be inspected visually or optically for surface defects. Subsurface and internal defects are investigated using various *nondestructive* techniques. In *destructive* testing, test specimens are removed from various sections of a casting to test for strength, ductility, and other mechanical properties, and to determine the presence and location of porosity and any other defects.[4]

Pressure tightness of cast components (valves, pumps, and pipes) is usually determined by sealing the openings in the casting and pressurizing it with water, oil, or air. (Because air is compressible, its use is dangerous in such tests because of the possibility of a sudden explosion due to a major flaw in the casting.) For extreme leak tightness requirements, pressurized helium or specially scented gases with detectors (sniffers) are used. The casting is then inspected for leaks while the pressure is maintained. Unacceptable or defective castings are re-melted for reprocessing.

Because of the major economic impact, the types of defects present in castings and their causes must be investigated. Control of all stages during casting, from mold preparation to the removal of castings from molds or dies, is important in maintaining good quality.

Notes

[1] Customarily, casting, forging, welding etc. are referred to as forming processes which

generally use molds, dies or other tooling to force the molten metal or metal blanks shaped into final parts. Machining (including turning, milling, shaping, grinding and drilling etc.), however, means remove the unwanted material from the blank surface in the form of chips with cutting tools or other physical and chemical means.

句意：人们习惯上将铸、锻、焊称为成形加工。这种工艺通常要采用模具和某些工艺装备迫使熔融金属或者毛坯成形为最终零件。而机械加工（包括车削、铣削、刨削、磨削和钻削等）是采用刀具或其他物理、化学手段，将毛坯表面不需要的部分以切屑的形式去除掉。

[2] The sequences involved in investment casting are shown in Fig. 3.2. The pattern is made by injecting molten wax or plastic into a metal die in the shape of the pattern. The pattern is then dipped into a slurry of refractory material such as very fine silica and binders, including water, ethyl silicate, and acids. After this initial coating has dried, the pattern is coated repeatedly to increase its thickness.

句意：熔模铸造的操作工序如图 3.2 所示。母模是通过将熔融的蜡或塑料注入金属模具中成形的。然后，将其浸入由耐火材料如微细石英粉和包含有水、硅酸乙酯和酸的黏合剂组成的浆料中。当这层初始涂挂材料干燥后，母模再经多次涂覆耐火浆料使涂层达到一定厚度。

[3] The expendable-pattern casting process uses a polystyrene pattern, which evaporates upon contact with molten metal to form a cavity for the casting. The process is also known as evaporative-pattern or lost-pattern casting, and under the trade name Full-Mold process.

句意：发泡模铸造采用（膨胀发泡的）聚苯乙烯作为母模，这种聚苯乙烯被熔融金属蒸发后在模型中留下一个型腔来生产铸件。因此，该工艺又称为气化模或消失模铸造，其商业名称又称为实型铸造。

[4] In destructive testing, test specimens are removed from various sections of a casting to test for strength, ductility, and other mechanical properties, and to determine the presence and location of porosity and any other defects.

句意：在破坏性试验中，试样从铸件的各个断面取得并进行强度、延展性及其他机械性能的检测，从而确定产生缩松和其他任何缺陷的位置所在。

Glossary

ablation   *n*. 热蚀
blank   *n*. 毛坯，坯料
centrifugal casting   离心铸造

cheek　*n.* 中（砂）箱
cope　*n.* 上箱
depolymerize　*v.*（使高分子化合物）解聚
drag　*n.* 下箱
engine block　发动机缸体
ethylsilicate　*n.* 硅酸乙酯
evaporative pattern　气化模
expandable pattern　（聚苯乙烯）发泡模
flask　*n.*（铸造）型箱
gate　*n.* 内浇口
investment casting　熔模铸造
lettering　*n.*（做成的、雕刻的）文字
lost-pattern casting　消失模铸造
nondestructive　*adj.* 非破坏性的
pattern　*n.*（铸造用）模型，母模（其他词义有模式、图纹、图案等）
pouring basin　浇口杯
pressure tightness　（容器的）密封性
refractory　*adj.* 难熔的，耐火的
riser　*n.* 冒口
runner　*n.* 横浇口
scented　*adj.* 有气味的
slurry　*n.* 浆料泥浆
sniffer　*n.* 嗅探器
sprue　*n.* 直浇口
vent　*n.* 通气孔

Unit 4　Forging and Die

4.1　Introduction

Forging is a process in which the workpiece is shaped by compressive forces applied through various dies and tools. It is one of the oldest metalworking operations, dating back at least to 4000 B.C.—perhaps as far back as 8000 B.C. Forging was first used to make jewelry, coins, and various implements by hammering metal with tools made of stone.

Simple forging operations can be performed with a heavy hand hammer and an anvil, as was traditionally done by blacksmiths. Most forgings, however, require a set of dies and such equipment as a press or a forging hammer. Unlike rolling operations, which generally produce continuous plates, sheets, strip, or various structural cross-sections, forging operations produce discrete parts.

Typical forged products are bolts and rivets, connecting rods, shafts for turbines, gears, hand tools, and structural components for machinery, aircraft, railroads, and a variety of other transportation equipment.

Metal flow and grain structure can be controlled, so forged parts have good strength and toughness, they can be used reliably for highly stressed and critical applications. Forging may be done at room temperature (cold forging) or at elevated temperatures (warm or hot forging).[1]

Because of the higher strength of the material, cold forging requires greater forces, and the workpiece materials must have sufficient ductility at room temperature. Cold-forged parts have good surface finish and dimensional accuracy. Hot forging requires smaller forces, but it produces lower dimensional accuracy and surface finish than cold-forged parts.

Forgings generally require additional finishing operations, such as heat treatment, to modify properties, and then machining to obtain accurate finished dimensions. These operations can be minimized by precision forging, which is an important example of the trend toward net-shape or near-net shape forming processes. This trend significantly reduces the number of operations required, and hence the manufacturing cost to make the final product.

4.2　Open-die Forging

Open-die forging is the simplest forging process. Although most open-die forgings generally weigh 15—500 kg (30—1000 lb), forgings as heavy as 300 tons can be made. Sizes may range from very small parts up to shafts about 23 m long (in the case of ship propellers).

The open-die forging process can be depicted by a solid workpiece placed between two flat dies and reduced in height by compressing it. This process is also called upsetting or flat-die forging.[2] The die surfaces in open-die forging may have simple cavities, to produce relatively simple forgings. Because constancy of volume is maintained, any reduction in height increases the diameter of the forged part.

4.3 Impression-die and Closed-die Forging

In impression-die forging, the workpiece acquires the shape of the die cavities (impressions) while being forged between two shaped dies (Fig.4.1). Note that some of the material flows outward and forms a flash. The flash has a significant role in the flow of material in impression-die forging. The thin flash cools rapidly, and because of its frictional resistance, it subjects the material in the die cavity to high pressures, thereby encouraging the filling of the die cavity.[3]

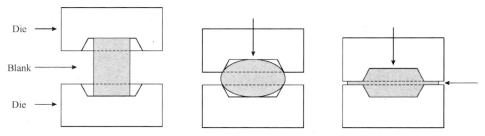

Fig.4.1 Stages in impression-die forging of a solid round billet

The blank is placed on the lower die and, as the upper die begins to descend, the blank's shape gradually changes, as is shown for the forging of a connecting rod in Fig.4.2(a). Preforming processes, such as fullering and edging [Fig.4.2(b) and Fig.4.2(c)], are used to distribute the material into various regions of the blank, much as in shaping dough to make pastry. In fullering, material is distributed away from an area; in edging, it is gathered in to a localized area.[4] The part is then formed into the rough shape of a connecting rod by a process called blocking, using *blocker dies*. The final operation is the finishing of the forging in impression dies that give the forging its final shape. The examples shown in Fig.4.1 and Fig.4.2(a) are also referred to as closed-die forgings. However, in true closed-die or flashless forging, flash does not form and the workpiece completely fills the die cavity. Accurate control of the volume of material and proper die design are essential in order to obtain a closed-die forging of the desired dimensions and tolerances. Undersize blanks prevent the complete filling of the die cavity; conversely, oversize blanks generate excessive pressures and may cause dies to fail prematurely or to jam.

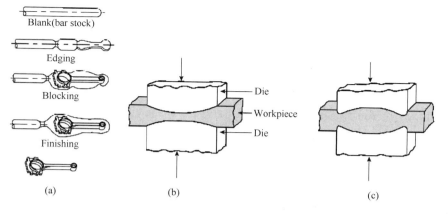

Fig.4.2 (a) Stages in forging a connecting rod; (b) Fullering operations; (c) Edging operations

4.4 Precision Forging

For economic reasons the trend in forging operations today is toward greater *precision*, which reduces the number of additional finishing operations. Operations in which the part formed is close to the final dimensions of the desired component are known as near-net-shape or net-shape forging. In such a process, there is little excess material on the forged part, and it is subsequently removed (generally by trimming or grinding).

In precision forging, special dies produce parts having greater accuracies than those from impression-die forging and requiring much less machining. The process requires higher-capacity equipment, because of the greater forces required to obtain fine details on the part. Because of the relatively low forging loads and temperatures that they require, aluminum and magnesium alloys are particularly suitable for precision forging; also, little die wear takes place and the surface finish is good. Steels and titanium can also be precision-forged. Typical precision-forged products are gears, connecting rods, housings, and turbine blades.

Coining. Coining essentially is a closed-die forging process typically used in minting coins, medallions, and jewelry; the slug is coined in a completely closed die cavity.[5] In order to produce fine details, the pressures required can be as high as five or six times the strength of the material. On some parts, several coining operations may be required. Lubricants cannot be applied in coining, because they can become entrapped in the die cavities and, being incompressible, prevent the full reproduction of die-surface details.

The coining process is also used with forgings and with other products, to improve surface finish and to impart the desired dimensional accuracy. This process, called sizing, involves high pressures, with little change in part shape during sizing. *Marking* of parts with letters and numbers can be done rapidly by a process similar to coining.

4.5 Die Manufacturing Methods

Various manufacturing methods, either singly or in combination, are used in making dies. These processes include casting, forging, machining, grinding, and electrical and electrochemical methods of die sinking. Dies are usually heat treated for greater hardness and wear resistance. If necessary, their surface profile and finish are improved by finish grinding and polishing, either by hand or by programmable industrial robots.

The choice of a die manufacturing method depends on the particular operation in which the die is to be used and on its size and shape. Cost often dictates the process selected, because tool and die costs can be significant in manufacturing operations. For example, a set of dies for automotive body panels can cost \$2 million. Even small and relatively simple dies can cost hundreds of dollars. On the other hand, because a large number of parts are usually made from the same die, die cost per piece is generally only a small portion of a part's manufacturing cost.

Dies may be classified as male and female; they may also be classified by their size. Small dies generally are those that have a surface area of 10^3—10^4 mm^2, whereas large dies have surface areas of 1 m^2 and larger, such as those used for pressworking automotive body panels.

Dies of various sizes and shapes can be cast from steels, cast irons, and nonferrous alloys. The processes used range from sand casting (for large dies weighing many tons) to shell molding (for small dies). Several die materials, such as tool and die steels, high-speed steels, and carbides, are often used to make dies. Cast steels are generally preferred for dies for large workpieces, because of their strength and toughness and the ease with which their composition, grain size, and properties can be controlled and modified.

Most commonly, dies are machined from forged die blocks by processes such as milling, turning, grinding, and electrical and electrochemical machining. Typically, a die for hot working operations is machined by milling on computer-controlled NC machines. Machining can be difficult for high-strength and wear-resistant die materials that are hard or heat treated.

These operations can be time consuming. As a result, nontraditional machining processes are used extensively, particularly for small- or medium-sized dies. These processes are generally faster and more economical, and the dies usually do not require additional finishing. Diamond dies for drawing fine wire are manufactured by producing holes with a thin rotating needle coated with diamond dust, using oil as a lubricant.

For improved hardness, wear resistance, and strength, die steels are usually heat treated. Improper heat treatment is one of the most common causes of die failure. Heat treatment may distort dies through the action of micro-structural changes and of uneven thermal cycling. Particularly important are the condition and composition of the die surfaces.

After heat treatment, dies are subjected to finishing operations, such as grinding, polishing, and chemical and electrical processes, to obtain the desired surface finish and dimensional accuracy. The grinding process, if not controlled properly, can cause surface damage from excessive heat and can induce harmful residual tensile stresses on the surface of the die, which will reduce its fatigue life.[6] Scratches on a die's surface can act as stress raisers. Likewise, commonly used die-making processes, such as electrical-discharge machining (EDM), can cause surface damage and cracks, unless the process parameters are carefully controlled.

Notes

[1] Metal flow and grain structure can be controlled, so forged parts have good strength and toughness, they can be used reliably for highly stressed and critical applications. Forging may be done at room temperature (cold forging) or at elevated temperatures (warm or hot forging).

句意：由于锻造可以对金属的流动和晶粒结构进行控制，因此锻件具有很好的强度和韧性，能可靠地用于承受较大应力和要求严格的工作环境。锻造可以在室温（冷锻）或高温（温锻或热锻）下进行。①

[2] The open-die forging process can be depicted by a solid workpiece placed between two flat dies and reduced in height by compressing it. This process is also called upsetting or flat-die forging.

句意：自由锻的操作过程是将工件放在两个平模之间，在压力作用下减小其厚度，故又称为墩粗或平模锻。

[3] The flash has a significant role in the flow of material in impression-die forging. The thin flash cools rapidly, and because of its frictional resistance, it subjects the material in the die cavity to high pressures, thereby encouraging the filling of the die cavity.

句意：锻件的飞边对金属材料在模锻过程中的流动有重要的影响。飞边较薄，冷却较快，造成摩擦阻力，使模腔中的金属材料由于无法继续向外流动而受到很高的压力，有利于它在模腔中的充填。

[4] Preforming processes, such as fullering and edging [Figs.4.2(b) and (c)], are used to distribute the material into various regions of the blank, much as in shaping dough to make pastry. In fullering, material is distributed away from an area; in edging, it is gathered in to a localized area.

---

① 这里有一点要说明：金属的再结晶温度是其熔点（热力学温度）的 0.4~0.6，在该温度以上锻造属于热锻，因此铅即使在室温下加工也算是热锻。——编者注

句意：（锻件毛坯的）预成形，如压肩和卡压（见图 4.2(b)、(c)）是对毛坯材料在不同区域进行分配，如我们在制作糕点时首先要将面团揉捏成初始的形状一样。卡压是使坯料从某处向两端分配，而压肩则是使坯料集中在某个局部区域。

[5] Coining essentially is a closed-die forging process typically used in minting coins, medallions, and jewelry; the slug is coined in a completely closed die cavity.

句意：精压是一种典型的闭式模锻成形工艺，常用于硬币、徽章和首饰的制作；小片的金属坯料在完全闭合的模腔中成形。

[6] The grinding process, if not controlled properly, can cause surface damage from excessive heat and can induce harmful tensile residual stresses on the surface of the die, which will reduce its fatigue life.

句意：对于磨削，如果操作不当会因为过热造成模具表面损伤，在模具表面会产生有害的残余拉伸应力，从而降低其疲劳寿命。

Glossary

automotive body panel　汽车车身覆盖件
closed-die forging　闭式模锻造
coining　*n.* 精压
die sinking　仿形制模（用电火花或电化学方法加工凹模）
edging　*n.* 压肩
electrical-discharge machining (EDM)　电火花加工
female die　阴模
flash　*n.* （锻件）飞边
flashless forging　无飞边锻造
flat die　平模
fullering　*n.* 卡压
impression-die forging　模锻
male die　阳模
medallion　*n.* 徽章
near-net shape forming　近净成形
open-die forging　自由锻
sizing　*n.* 精整
trimming　*n.* 去毛刺，修边
upsetting　*n.* 墩粗
warm forging　温锻

Unit 5　Conventional Machining Processes

5.1　Introduction

Conventional machining is the group of machining operations that use single- or multi-point tools to remove material in the form of chips. Metal cutting involves removing metal through machining operations. Machining traditionally takes place on lathes, drill presses, and milling machines with the use of various cutting tools. Most machining has very low set-up cost compared with forming, molding, and casting processes. Machining is necessary where tight tolerances on dimensions and finishes are required.

5.2　Turning and Lathe

Turning is one of the most common of metal cutting operations. In turning, a workpiece is rotated about its axis as single-point cutting tools are fed into it, cutting away excess material and creating the desired cylindrical surface. Turning can occur on both external and internal surfaces to produce an axially-symmetrical contoured part. Parts ranging from pocket watch components to large diameter marine propeller shafts can be turned on a lathe.

Apart from turning, several other operations can also be performed on a lathe.

Boring and internal turning. Boring and internal turning are performed on the internal surfaces by a boring bar or suitable internal cutting tools. If the initial workpiece is solid, a drilling operation must be performed first.[1] The drilling tool is held in the tailstock, and the latter is then fed against the workpiece. When boring is done in a lathe, the work is usually held in a chuck or on a face plate. Holes may be bored straight, tapered, or to irregular contours. Boring is essentially internal turning while feeding the tool parallel to the rotation axis of the workpiece.

Facing. Facing is the producing of a flat surface as the result of a tool's being fed across the end of the rotating workpiece. Unless the work is held on a mandrel, if both ends of the work are to be faced, it must be turned around after the first end is completed and then the facing operation repeated.[2] The cutting speed should be determined from the largest diameter of the surface to be faced. Facing may be done either from the outside inward or from the center outward. In either case, the point of the tool must be set exactly at the height of the center of rotation. Because the cutting force tends to push the tool away from the work, it is usually desirable to clamp the carriage to the lathe bed during each facing cut to prevent it from moving slightly and thus producing a surface that is not flat. In the facing of casting or other materials that have a hard surface, the depth of the first cut should be sufficient to penetrate the hard material to avoid excessive tool wear.

Parting. Parting is the operation by which one section of a workpiece is severed from the remainder by means of a cutoff tool. Because cutting tools are quite thin and must have considerable overhang, this process is less accurate and more difficult. The tool should be set exactly at the height of the axis of rotation, be kept sharp, have proper clearance angles, and be fed into the workpiece at a proper and uniform feed rate.

Threading. Threading can be considered as turning since the path to be travelled by the cutting tool is helical. However, there are some major differences between turning and threading. While in turning, the interest is in generating a smooth cylindrical surface, in threading the interest is in cutting a helical thread of a given form and depth which can be calculated from the formulae. There are two basic requirements for thread cutting. An accurately shaped and properly mounted tool is needed because thread cutting is a form-cutting operation. The resulting thread profile is determined by the shape of the tool and its position relative to the workpiece. The second requirement is that the tool must move longitudinally in a specific relationship to the rotation of the workpiece, because this determines the lead of the thread. This requirement is met through the use of the lead screw and the split unit, which provide positive motion of the carriage relative to the rotation of the spindle.

Many types of lathes are used for production turning. According to purposes and construction, lathe-type machine tools can be classified as follows:

1. Engine lathes;
2. Vertical lathes;
3. Turret lathes;
4. Single- or multiple-spindle automatic or semi-automatic lathes;
5. Contouring lathes;
6. Universal lathes;
7. Special-purpose lathes such as crankshaft lathes, camshaft lathes, car wheel lathes and backing-off lathes, etc.

The engine lathe is the most representative member of the lathe family and is the most widely used, so there is a description of each of the main elements of an engine lathe, which is shown in Fig.5.1.

Lathe bed is the foundation of the engine lathe, which is a heavy, rugged casting made to support the working parts of the lathe. The size and mass of the bed give the rigidity necessary for accurate engineering tolerances required in manufacturing. On top of the bed are machined slideways that guide and align the carriage and tailstock, as they move from one end of the lathe to the other.

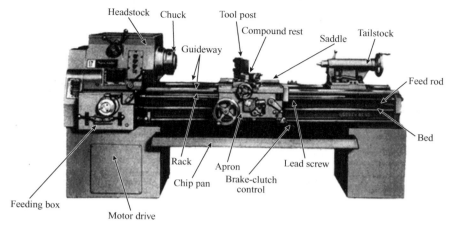

Fig.5.1 The engine lathe

Headstock is clamped atop the bed at the left-hand end of the lathe and contains the motor that drives the spindle whose axis is parallel to the guideways through a series of gears housed within the gearbox. The function of gearbox is to generate a number of different spindle speeds. A spindle gear is mounted on the rear of the spindle to transmit power through the change gears to the feeding box that distributes the power to the lead screw for threading or to the feed rod for turning.[3]

The spindle has a through hole extending lengthwise through which bar stocks can be fed if continuous production is used. The hole can hold a plain lathe center by its tapered inner surface and mount a chuck, a face plate or collet by its threaded outer surface.

Carriage assembly is actually an H-shaped block that sits across the guideways and in front of the lathe bed. The function of the carriage is to carry and move the cutting tool longitudinally. It can be moved by hand or by power and can be clamped into position with a locking nut. The carriage is composed of the cross slide, compound rest, tool saddle, and apron.

The cross slide is mounted on the dovetail guideways on the top of the saddle and is moved back and forth at 90° to the axis of the lathe by the cross slide lead screw. The lead screw can be hand or power activated.

The compound rest is mounted on the cross slide and can be swiveled and clamped at any angle in a horizontal plane. The compound is typically used for cutting chamfers or tapers, but must also be used when cutting threads. The compound rest can only be fed by hand. There is no power to the compound rest. The cutting tool and tool holder are secured in the tool post which is mounted directly to the compound rest.

The tool saddle is an H-shaped casting mounted on top of the guideways and houses the cross slide and compound rest. It makes possible longitudinal, cross and angular feeding of the tool bit.

The apron is attached to the front of the carriage and contains the gears and feed clutches which transmit motion from the feed rod or lead screw to the carriage and cross slide. When cutting screw threads, power is provided to the gearbox of the apron by the lead screw. In all other turning operations, it is the feed rod that drives the carriage.

Tailstock is composed of a low base and the movable part of the tail-stock proper, the transverse adjustments being made with a cross screw furnished with a square head. The two parts are held together by the holding-down bolts which secure the tailstock to the bed. The tailstock is located on the opposite end of the lathe from the headstock. It supports one end of the work when machining between centers, supports long pieces held in the chuck, and holds various forms of cutting tools, such as drills, reamers, and taps.

5.3 Milling and Milling Machine

Milling is the process of cutting away material by feeding a workpiece past a rotating multiple-tooth cutter. The cutting action of several teeth around the milling cutter provides a fast and smooth method of machining. The machined surface may be flat, angular, or curved. The surface may also be milled to any combination of shapes. The machine for holding the workpiece, rotating the cutter, and feeding it is known as the milling machine.

There are several milling operations as follows:

Peripheral milling. In peripheral (or slab) milling, the milled surface is generated by teeth located on the periphery of the cutter body.[4] The axis of cutter rotation is generally in a plane parallel to the workpiece surface to be machined [shown in Fig.5.2(a)].

Face milling. In face milling, the cutter is mounted on a spindle having an axis of rotation perpendicular to the workpiece surface. The milled surface results from the action of cutting edges located on the periphery and face of the cutter [shown in Fig.5.2(b)].

End milling. The cutter in end milling generally rotates around an axis vertical to the workpiece. It can be tilted to machine tapered surfaces. Cutting teeth are located on both the end face of the cutter and the periphery of the cutter body [shown in Fig.5.2(c)].

Milling machines are basically classified as vertical or horizontal according to the orientation of their spindle axes. A vertical milling machine spindle axis is vertical and the horizontal milling machine spindle axis is horizontal. In addition, the vertical milling machine has a machine table that moves perpendicular to the spindle axis of rotation and the horizontal milling machine has a worktable that moves parallel to the spindle axis of rotation. These machines are also classified as column and knee type, ram-type, manufacturing or bed type, and planer-type.

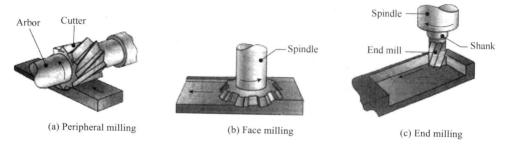

(a) Peripheral milling (b) Face milling (c) End milling

Fig.5.2 Some basic types of milling operations

Used for general purpose milling operations, column and knee type milling machines are the most common milling machines. The spindle to which the milling cutter is clutched may be horizontal (slab milling) or vertical (face and end milling shown in Fig.5.3). The basic components are:

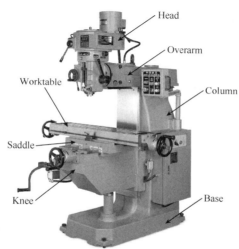

Fig.5.3 The vertical milling machine

Column is the main casting which includes the base and supports all other parts of the machine. An oil reservoir and a pump in the column keep the spindle lubricated. The column rests on a base that contains a coolant reservoir and a pump that can be used when performing any machining operation that requires a coolant.

Knee is the casting that supports the table and the saddle that gives the table vertical movements for adjusting the depth of cut.[5] The feed change gearing is enclosed within the knee. It is supported and can be adjusted by the elevating screw. The knee is fastened to the column by dovetail guideways. The lever can be raised or lowered either by hand or power feed. The hand feed is usually used to take the depth of cut or to position the work, and the power feed to move the work during the machining operation.

Saddle supports the table and can move transversely, which slides on a horizontal dovetail, parallel to the axis of the spindle, on the knee. The swivel table (on universal machines only) is attached to the saddle and can be swiveled approximately 45° in either direction.

Power feed mechanism is contained in the knee and controls the longitudinal, transverse (in and out) and vertical feeds. The desired feed rate can be obtained on the machine by positioning the feed selection levers as indicated on the feed selection plates. On some universal knee and column milling machines the feed is obtained by turning the speed selection handle until the desired rate of feed is indicated on the feed dial. Most milling machines have a rapid traverse lever that can be engaged when a temporary increase in speed of the longitudinal, transversal or vertical feeds is required. For example, this lever would be engaged when positioning or aligning the work.

Worktable is the rectangular casting located on top of the saddle. It contains several T-slots for fastening the work or workholding devices. The table can be moved longitudinally with respect to the saddle by hand or by power. To move the table by hand, engage and turn the longitudinal hand crank. To move it by power, engage the longitudinal directional feed control lever. The longitudinal directional control lever can be positioned to the left, to the right, or in the center. Place the end of the directional feed control lever to the left to feed the table to the left. Place it to the right to feed the table to the right. Place it in the center position to disengage the power feed, or to feed the table by hand.

Head contains the spindle and cutter holders. In vertical machines the head may be fixed or vertically adjustable. The spindle holds and drives various cutting tools. It is a shaft, mounted on bearings supported by the column. The spindle is driven by an electric motor through a train of gears, all mounted within the column. The front end of the spindle, which is near the table, has an internal taper machined on it. The internal taper ($3\frac{1}{2}$ inches per foot) permits mounting tapered-shank cutter holders and cutter arbors. Two keys, located on the face of the spindle, provide a positive drive for the cutter holder, or arbor. The holder or arbor is secured in the spindle by a draw bolt and jam nut. Large face mills are sometimes mounted directly to the spindle nose.

Overarm is the horizontal beam to which the arbor support is fastened. The overarm may be a single casting that slides in the dovetail guideways on the top of the column. It may consist of one or two cylindrical bars that slide through the holes in the column. On some machines to position the overarm, first unclamp the locknuts and then extend the overarm by turning a crank. On others, the overarm is moved by merely pushing on it. The overarm should only be extended far enough to so position the arbor support as to make the setup as rigid as possible. To place the arbor supports on an overarm, extend one of the bars approximately 1-inch farther than the other bar.

5.4 Drilling and Drill Press

Drill can be defined as a rotary end cutting tool having one or more cutting lips, and having one or more helical or straight flutes for the passage of chips and the admission of a cutting fluid. There are several hole-making operations carried out on the drill press as follows:

Drilling. Drilling involves selecting the proper twist drill or cutter for the job, properly installing the drill into the machine spindle, setting the speed and feed, drilling a smaller pilot hole, and drilling the hole to specifications within the prescribed tolerance. Drilled holes are always slightly oversized, or slightly larger than the diameter of the drill's original designation.

Reaming. Reaming can be performed on a drilling machine. It is difficult, if not impossible, to drill a hole to an exact standard diameter. When great accuracy is required, the holes are first drilled slightly undersized and then reamed to size. Reaming can be done on a drilling machine by using a hand reamer or using a machine reamer. When you must drill and ream a hole, it is best if the setup is not changed.

Tapping. Tapping is cutting a thread in a drilled hole. Tapping is accomplished on the drilling machine by selecting and drilling the tap drill size, then using the drilling machine chuck to hold and align the tap while it is turned by hand. The drilling machine is not a tapping machine, so it should not be used to power tap. To avoid breaking taps, ensure the tap aligns with the center axis of the hole, keep tap flutes clean to avoid jamming, and clean chips out of the bottom of the hole before attempting to tap.

Counterboring. Counterboring is the process of using a counterbore to enlarge the upper end of a hole to a predetermined depth and machine a recess at that depth.[6] Counterbored holes are primarily used to recess socket head cap screws and similar bolt heads slightly below the surface.

Countersinking. Countersinking is an operation in which a cone-shaped enlargement is cut at the top of a hole to form a recess below the surface. A conical cutting tool is used to produce this chamfer. When countersinking, the cutter must be properly aligned with the existing hole. Countersinking is useful in removing burrs from edges of holes, as well as accommodating the countersunk screw head.

Spot-facing. Spot-facing is basically the same as counterboring, using the same tool, speed, feed, and lubricant. Spot-facing is the smoothing off and squaring of a rough or curved surface around a hole to permit level seating of washers, nuts, or bolt heads. The operation of spot-facing is slightly different in that the spot-facing is usually done above a surface or on a curved surface.[7] When rough surfaces, castings, and curved surfaces are not at right angles to the cutting tool, great strain may occur on the pilot hole and counterbore, which can lead to broken tools. Both counterboring and spot-facing can be accomplished with standard counterbore cutters.

Boring. Boring is conducted when a straight and smooth hole is needed occasionally which is too large or odd sized for drills or reamers. A boring tool can be inserted into the drilling machine and bore any size hole into which the tool holder will fit. A boring bar with a tool bit installed is used for boring on the larger drilling machines. To bore accurately, the setup must be rigid, machine must be sturdy, and power feed must be used. Boring is not recommended for hand-feed drilling machines. Hand feed is not smooth enough for boring and can be dangerous, because the tool bit could catch the workpiece and throw it back to the operator.

There are many different types or configurations of drill presses, but most drill presses will fall into four broad categories: sensitive bench type, upright type, radial type, and special purpose type.[8]

Rigid and accurate construction of drilling machines is important to obtain proper results with the various cutting tools used. The sensitive drilling machine construction features are discussed in this section because its features are common to most other drilling machines (Fig. 5.4).

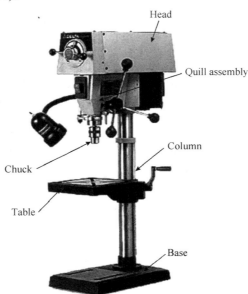

Fig.5.4   The bench type drilling press

Base is the main supporting member of the machine. It is heavy gray iron or ductile iron casting with slots to support and hold work that is too large for the table.

Column may be made of gray cast iron or ductile iron for larger machines, or steel tubing for smaller bench drill presses. It supports the table and the head of the drilling machine. The outer surface is machined to function as a precision way of aligning the spindle with the table.

Chuck is used to clamp and fix the drill through a cone-surface locking mechanism.

Table can be adjusted up or down the column to the proper height. It can also be swiveled around the column to the desired working position. Most worktables have slots and holes for mounting vises and other workholding accessories. Some tables are semi-universal, meaning that they can be swiveled about the horizontal axis.

Head houses the spindle, quill, pulleys, motor and feed mechanism. The V-belt from the motor drives a pulley in the front part of the head, which in turn drives the spindle. The spindle turns the drill. Speeds on a stepped V pulley drive are changed by changing the position of the V-belt. Speeds on a variable-speed drive mechanism are changed by a hand wheel on the head. The spindle must be revolving when this is done.

Quill assembly makes it possible to feed or withdraw the cutting tool from the work. The spindle rotates within the quill on bearings. The quill moves vertically by means of a rack and pinion. Located on the lower end of the spindle is either a Morse tapered hole or a threaded stub where the drill chuck is mounted. For drilling larger holes, the drill chuck is removed and Morse tapered cutting tools are mounted.

Notes

[1] If the initial workpiece is solid, a drilling operation must be performed first.

句意：如果工件毛坯是实心的，首先要钻孔。

[2] Unless the work is held on a mandrel, if both ends of the work are to be faced, it must be turned around after the first end is completed and the facing operation repeated.

句意：除非工件固定在心轴上，如果工件两端都要车端面，就必须在一端加工完成后，将工件倒过来，重复进行车端面加工。

[3] A spindle gear is mounted on the rear of the spindle to transmit power through the change gears to the feeding box that distributes the power to the lead screw for threading or to the feed rod for turning.

句意：主轴齿轮安装在主轴尾部，通过挂轮把动力传递到进给箱。如果是车螺纹，进给箱就将动力分配到丝杠上；如果是车削，就将动力分配到光杠上。

[4] In peripheral (or slab) milling, the milled surface is generated by teeth located on the periphery of the cutter body.

句意：在圆周（平面）铣削中，铣削表面是由分布在刀体周围的刀齿加工完成的。

[5] Knee is the casting that supports the table and the saddle that gives the table vertical movements for adjusting the depth of cut.

句意：升降台是一个铸件，用来支撑工作台和床鞍，它可以实现工作台的垂直移动以调整切深。

[6] Counterboring is the process of using a counterbore to enlarge the upper end of a hole to a predetermined depth and machine a recess at that depth.

句意：锪沉头孔加工就是利用平底扩孔钻将孔的上端部扩至预定深度，从而加工出一个凹窝（以便螺栓头的埋入）。

[7] The operation of spot facing is slightly different in that the spot facing is usually done above a surface or on a curved surface.

句意：锪凸台加工稍微有所不同，它通常是在平面上或曲面上进行的。

[8] There are many different types or configurations of drill presses, but most drill presses will fall into four broad categories: sensitive bench type, upright type, radial type, and special purpose type.

句意：钻床有很多种类型，但多数钻床都分成四大类：台式钻床、立式钻床、摇臂式钻床和专用钻床。

Glossary

approximately　*adv.* 近似地，大约
apron　*n.* 拖板箱，溜板箱
arbor　*n.* 刀杆，刀轴；柄轴
axially　*adv.* 沿轴向地
boring　*n.* 镗孔，镗削
carriage　*n.* 大刀架；拖板
chuck　*n.* 卡盘，钻轧头
clamp　*n.* 夹子，钳子　*v.* 夹住，夹紧
compound rest　小刀架
conical　*adj.* 圆锥（形，体）的
contour　*n.* 轮廓，外形
coolant　*n.* 冷却剂
counterboring　*n.* 锪沉头孔
countersinking　*n.* 锪锥孔
cross slide　横向滑板，横向拖板
cutoff tool　切断刀
cylindrical　*adj.* 圆柱的

dovetail　　*n.* 燕尾榫，燕尾接合　*v.* 吻合
draw bolt　　牵引螺栓；接合螺栓
drill press　　钻床
ductile　　*adj.* 易拉长的，易变形的，可塑的
end milling　　立铣
face milling　　端面铣削
facing　　*n.* 端面车削，平面加工
feed rod　　光杠
feeding box　　进给箱
gearbox　　*n.* 齿轮箱，变速箱
headstock　　*n.* 主轴箱
jam nut　　扁螺母；止动螺母
lathe　　*n.* 车床　*v.* 用车床加工
lead screw　　丝杠
lengthwise　　*adj.* 纵长的　*adv.* 纵长地
lever　　*n.* 杠杆，推杆　*v.* 撬
locknut　　*n.* 防松螺母
longitudinal　　*adj.* 纵向的，经线的
longitudinally　　*adv.* 纵向地；经线地
overarm　　*n.* 横臂，横杆
parting　　*n.* 切断；分开，分割，剖切
penetrate　　*v.* 穿透；渗透
peripheral milling　　圆周铣削
perpendicular　　*adj.* 垂直的，正交的　*n.* 垂直线，垂直度
quill assembly　　钻套组件
reamer　　*n.* 铰刀
reaming　　*n.* 铰孔
recess　　*n.* 凹进部分，凹坑　*v.* 使凹进
remainder　　*n.* 剩余（物）；余数
reservoir　　*n.* 油箱；容器；储液槽
rigidity　　*n.* 刚性，刚度；硬度
shear　　*v.* 剪（切）
slab milling　　平面铣削，阔面铣
slideway　　*n.* 导轨，滑轨；滑槽
slot　　*n.* 槽，插槽；缝，裂缝；切口　*v.* 在……上开槽，在……上开狭长的孔
socket　　*n.* 插孔；插口；槽；座

spindle　*n.* 主轴
spindle nose　主轴端部，轴头
spotfacing　*n.* 锪凸台
swivel　*n.* 转环，转体　*v.*（使）转动，（使）旋转
symmetrical　*adj.* 对称的，匀称的
tailstock　*n.* 尾座，尾架
tap　*n.* 丝锥　*v.* 攻丝
taper　*n.* 坡度；锥度；锥形　*v.* 锥度加工；逐渐变细
tapping　*n.* 攻螺纹，攻螺丝
temporary　*adj.* 暂时的，临时的；短暂的
thread　*n.* 螺纹（齿，丝，线）
tool saddle　刀架鞍板
transverse　*adj.* 横向的，横切的　*n.* 横轴
turning　*n.* 车削；旋转；弯曲；转向
twist drill　麻花钻
unclamp　*v.* 松开
uniform　*n.* 均匀；一致
vise　*n.* 老虎钳　*v.* 钳住，紧紧夹住

Unit 6　Nontraditional Machining Processes

6.1　Introduction

Traditional or conventional machining, such as turning, milling, and grinding etc., uses mechanical energy to shear metal against another substance to create holes or remove material. Nontraditional machining processes are defined as a group of processes that remove excess material by various techniques involving mechanical, thermal, electrical or chemical energy or combinations of these energies but do not use a sharp cutting tool as it is used in traditional manufacturing processes.[1]

Extremely hard and brittle materials are difficult to be machined by traditional machining processes. Using traditional methods to machine such materials means increased demand for time and energy and therefore increases in costs; in some cases traditional machining may not be feasible. Traditional machining also results in tool wear and loss of quality in the product owing to induced residual stresses during machining. Nontraditional machining processes, also called unconventional machining process or advanced manufacturing processes, are employed where traditional machining processes are not feasible, satisfactory or economical due to special reasons as outlined below:

1. Very hard fragile materials difficult to clamp for traditional machining;
2. When the workpiece is too flexible or slender;
3. When the shape of the part is too complex;
4. Parts without producing burrs or inducing residual stresses.

Traditional machining can be defined as a process using mechanical (motion) energy. Nontraditional machining utilizes other forms of energy; the three main forms of energy used in nontraditional machining processes are as follows:

1. Thermal energy;
2. Chemical energy;
3. Electrical energy.

Several types of nontraditional machining processes have been developed to meet extra required machining conditions. When these processes are employed properly, they offer many advantages over traditional machining processes. The common nontraditional machining processes are described in the following section.

6.2 Electrical Discharge Machining (EDM)

Electrical discharge machining (EDM) sometimes is colloquially referred to as *spark machining*, *spark eroding*, *burning*, *die sinking* or *wire erosion*. It is one of the most widely used nontraditional machining processes. The main attraction of EDM over traditional machining processes such as metal cutting using different tools and grinding is that this technique utilizes thermoelectric process to erode undesired materials from the workpiece by a series of rapidly recurring discrete electrical sparks between workpiece and electrode.[2]

The traditional machining processes rely on harder tool or abrasive material to remove the softer material whereas nontraditional machining processes such as EDM uses electrical spark or thermal energy to erode unwanted material in order to create desired shapes. So, the hardness of the material is no longer a dominating factor for EDM process.

EDM removes material by discharging an electrical current, normally stored in a capacitor bank, across a small gap between the tool (cathode) and the workpiece (anode) typically in the order of 50 volts/10amps. As shown in Fig.6.1, at the beginning of EDM operation, a high voltage is applied across the narrow gap between the electrode and the workpiece. This high voltage induces an electric field in the insulating dielectric that is present in the narrow gap between electrode and workpiece. This causes conducting particles suspended in the dielectric to concentrate at the points of strongest electrical field. When the potential difference between the electrode and the workpiece is sufficiently high, the dielectric breaks down and a transient spark discharges through the dielectric fluid, removing small amount of material from the workpiece surface. The volume of the material removed per spark discharge is typically in the range of 10^{-6} to 10^{-5} mm^3. The gap is only a few thousandths of an inch, which is maintained at a constant value by the servomechanism that actuates and controls the tool feed.

6.3 Chemical Machining (CM)

Chemical machining is a well known non-traditional machining process in which metal is removed from a workpiece by immersing it into a chemical solution. The process is the oldest of the nontraditional processes and has been used to produce pockets and contours and to remove materials from parts having a high strength-to-weight ratio. Moreover, the chemical machining method is widely used to produce micro-components for various industrial applications such as microelectromechanical systems (MEMS) and semiconductor industries.[3]

In CM material is removed from selected areas of workpiece by immersing it in a chemical reagents or etchants, such as acids and alkaline solutions. Material is removed by microscopic electrochemical cell action which occurs in corrosion or chemical dissolution of a metal. Special coatings called maskants protect areas from which the metal is not to be removed. This controlled chemical dissolution will simultaneously etch all exposed surfaces even though the penetration rates of the material removal may be only 0.0025—0.1 mm/min. The basic process takes many forms: chemical milling of pockets, contours, overall metal removal, chemical blanking for etching through thin sheets; photochemical machining (PCM) for etching by using of photosensitive resists in microelectronics; chemical or electrochemical polishing where weak chemical reagents are used (sometimes with remote electric assist) for polishing or deburring and chemical jet machining where a single chemically active jet is used. A schematic of chemical machining process is shown in Fig.6.2(a). Because the etchant attacks the material in both vertical and horizontal directions, undercuts may develop [as shown by the areas under the edges of the maskant in Fig.6.2(b)]. Typically, tolerances of ±10% of the material thickness can be maintained in chemical blanking. In order to improve the production rate, the bulk of the workpiece should be shaped by other processes (such as by machining) prior to chemical machining. Dimensional variations can occur because of size changes in workpiece due to humidity and temperature. This variation can be minimized by properly selecting etchants and controlling the environment in the part generation and the production area in the plant.

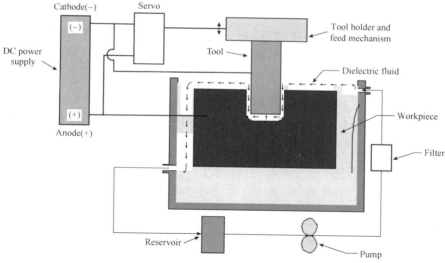

Fig.6.1 Basic electro-discharge system

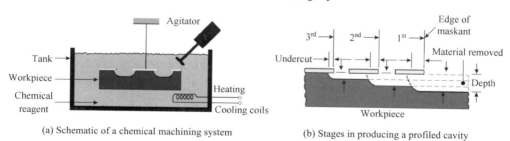

(a) Schematic of a chemical machining system   (b) Stages in producing a profiled cavity

Fig.6.2 Schematic illustration of chemical machining process

6.4 Electrochemical Machining (ECM)

Electrochemical metal removal is one of the more useful nontraditional machining processes. Although the application of electrolytic machining as a metal-working tool is relatively new, the basic principles are based on Faraday laws. Thus, electrochemical machining can be used to remove electrically conductive workpiece material through anodic dissolution. No mechanical or thermal energy is involved. This process is generally used to machine complex cavities and shapes in high-strength materials, particularly in the aerospace industry for the mass production of turbine blades, jet-engine parts, and nozzles, as well as in the automotive (engines castings and gears) and medical industries.[4] More recent applications of ECM include micromachining for the electronics industry.

ECM, shown in Fig.6.3, is a metal-removal process based on the principle of reverse electroplating. In this process, particles travel from the anodic material (workpiece) toward the cathodic material (machining tool). Metal removal is effected by a suitably shaped tool electrode, and the parts thus produced have the specified shape, dimensions, and surface finish. ECM forming is carried out so that the shape of the tool electrode is transferred onto, or duplicated in, the workpiece. The cavity produced is the female mating image of the tool shape. For high accuracy in shape duplication and high rates of metal removal, the process is operated at very high current densities of the order $10-100$ A/cm^2, at relative low voltage usually from 8 to 30 V, while maintaining a very narrow machining gap (of the order of 0.1 mm) by feeding the tool electrode with a feed rate from 0.1 to 20 mm/min. Dissolved material, gas, and heat are removed from the narrow machining gap by the flow of electrolyte pumped through the gap at a high velocity (5—50 m/s), so the current of electrolyte fluid carries away the deplated material before it has a chance to reach the machining tool.

Being a non-mechanical metal removal process, ECM is capable of machining any electrically conductive material with high stock removal rates regardless of their mechanical properties.[5] In particular, removal rate in ECM is independent of the hardness, toughness and other properties of the material being machined. The use of ECM is most warranted in the manufacturing of complex-shaped parts from materials that lend themselves poorly to machining by other, above all mechanical methods. There is no need to use a tool made of a harder material than the workpiece, and there is practically no tool wear. Since there is no contact between the tool and the work, ECM is the machining method of choice in the case of thin-walled, easily deformable components and also brittle materials likely to develop cracks in the surface layer.

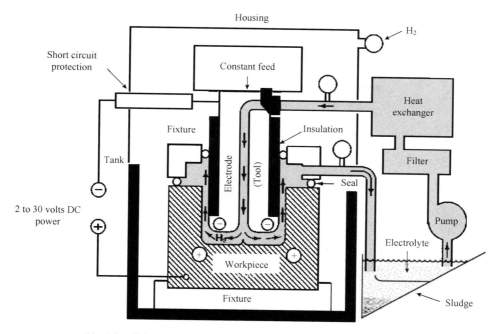

Fig.6.3 Schematic diagram of electrochemical machining process

6.5 Laser Beam Machining (LBM)

LASER is an acronym for Light Amplification by Stimulated Emission of Radiation. Although the laser is used as a light amplifier in some applications, its principal use is as an optical oscillator or transducer for converting electrical energy into a highly collimated beam of optical radiation.[6] The light energy emitted by the laser has several characteristics which distinguish it from other light sources: spectral purity, directivity and high focused power density.

Laser machining is the material removal process accomplished through laser and target material interactions. Generally speaking, these processes include laser drilling, laser cutting, laser welding, and laser grooving, marking or scribing.

Laser machining (Fig.6.4) is localized, non-contact machining and is almost reacting-force free.[7] This process can remove material in very small amount and is said to remove material "atom by atom". For this reason, the kerf in laser cutting is usually very narrow, the depth of laser drilling can be controlled to less than one micron per laser pulse and shallow permanent marks can be made with great flexibility. In this way material can be saved, which may be important for precious materials or for delicate structures in micro-fabrications. The ability of accurate control of material removal makes laser machining an important process in micro-fabrication and micro-electronics. Also laser cutting of sheet material with thickness less than 20 mm can be fast, flexible and of high quality, and large holes or any complex contours can be efficiently made through trepanning.

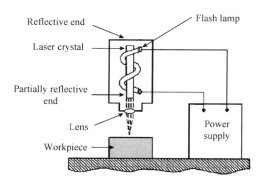

Fig.6.4　Schematic diagram of laser beam machining device

Heat affected zone (HAZ) in laser machining is relatively narrow and the re-solidified layer is of micron dimensions.[8] For this reason, the distortion in laser machining is negligible. LBM can be applied to any material that can properly absorb the laser irradiation. It is difficult to machine hard materials or brittle materials such as ceramics using traditional methods, laser is a good choice for solving such difficulties.

Laser cutting edges can be made smooth and clean, no further treatment is necessary. High aspect ratio holes with diameters impossible for other methods can be drilled using lasers. Small blind holes, grooves, surface texturing and marking can be achieved with high quality using LBM. Laser technology is in rapid progressing, so do laser machining processes. Dross adhesion and edge burr can be avoided, geometry accuracy can be accurately controlled. The machining quality is in constant progress with the rapid progress in laser technology.

6.6　Ultrasonic Machining (USM)

Ultrasonic machining offers a solution to the expanding need for machining brittle materials such as single crystals, glasses and polycrystalline ceramics, and for increasingly complex operations to provide intricate shapes and workpiece profiles. This machining process is non-thermal, non-chemical, creates no change in the microstructure, chemical or physical properties of the workpiece and offers virtually stress-free machined surfaces. It is therefore used extensively in machining hard and brittle materials that are difficult to cut by other traditional methods. The actual cutting is performed either by abrasive particles suspended in a fluid, or by a rotating diamond-plated tool. These variants are known respectively as stationary (conventional) ultrasonic machining and rotary ultrasonic machining (RUM).

Conventional ultrasonic machining accomplishes the removal of material by the abrading action of a grit-loaded slurry, circulating between the workpiece and a tool that is vibrated with small amplitude. The form tool itself does not abrade the workpiece; the vibrating tool excites the abrasive grains in the flushing fluid, causing them to gently and uniformly wear away the material, leaving a precise reverse form of the tool shape. The uniformity of the sonotrode-tool vibration limits the process to forming small shapes typically under 100 mm in diameter.

The USM system includes the sonotrode-tool assembly, the generator, the grit system and the operator controls. The *sonotrode* is a piece of metal or tool that is exposed to *ultrasonic* vibration, and then gives this vibratory energy in an element to excite the abrasive grains in the slurry.[9] A schematic representation of the USM set-up is shown in Fig.6.5. The sonotrode-tool assembly consists of a transducer, a booster and a sonotrode. The transducer converts the electrical pulses into vertical stroke. This vertical stroke is transferred to the booster, which may amplify or suppress the stroke amount. The modified stroke is then relayed to the sonotrode-tool assembly. The amplitude along the face of the tool typically falls in a 20 to 50 μm range. The vibration amplitude is usually equal to the diameter of the abrasive grit used.

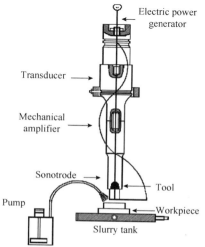

Fig.6.5 Schematic representation of USM apparatus

The grit system supplies a slurry of water and abrasive grit, usually silicon or boron carbide, to the cutting area. In addition to providing abrasive particles to the cut, the slurry also cools the sonotrode and removes particles and debris from the cutting area.

Notes

[1] Nontraditional manufacturing processes is defined as a group of processes that remove excess material by various techniques involving mechanical, thermal, electrical or chemical energy or combinations of these energies but do not use a sharp cutting tools as it needs to be used for traditional manufacturing processes.

句意：特种加工是指这样一组加工工艺，它们通过各种涉及机械能、热能、电能、化学能或及其组合形式的技术，而不使用传统加工所必需的尖锐刀具来去除工件表面的多余材料。

[2] The main attraction of EDM over traditional machining processes such as metal cutting using different tools and grinding is that this technique utilizes thermoelectric process to erode undesired materials from the workpiece by a series of rapidly recurring discrete electrical sparks between workpiece and electrode.

句意：相比于利用不同刀具进行金属切削和磨削的常规加工，电火花加工更为吸引人之处在于它利用工件和电极间的一系列快速重复产生的（脉冲）离散电火花所产生的热电作用，从工件表面通过电腐蚀去除多余的材料。

[3] The chemical machining method is widely used to produce micro-components for various industrial applications such as microelectromechanical systems (MEMS) and semiconductor industries.

句意：化学加工广泛用于为多种工业应用（如微机电系统和半导体行业）制造微型零件。

[4] This process is generally used to machine complex cavities and shapes in high-strength materials, particularly in the aerospace industry for the mass production of turbine blades, jet-engine parts, and nozzles, as well as in the automotive (engines castings and gears) and medical industries.

句意：这个加工过程一般用于在高强度材料上加工复杂型腔和形状，特别是在航空业中如加工涡轮机叶片、喷气发动机零件和喷嘴，以及在汽车业（发动机铸件和齿轮）和医疗卫生业中相关零件的加工。

[5] Being a non-mechanical metal removal process, ECM is capable of machining any electrically conductive material with high stock removal rates regardless of their mechanical properties.

句意：作为一种非机械式金属去除加工方法，ECM 可以以高切削量加工任何导电材料，而无须考虑材料的机械性能。

[6] LASER is an acronym for Light Amplification by Stimulated Emission of Radiation. Although the laser is used as a light amplifier in some applications, its principal use is as an optical oscillator or transducer for converting electrical energy into a highly collimated beam of optical radiation.

句意：LASER 是英文 Light Amplification by Stimulated Emission of Radiation 的缩写词。意思是"通过受激发射光放大"或"光的受激发射"，简称激光。虽然激光在某些应用场合可用来作为放大器，但它的主要用途是光激射振荡器，或者是作为将电能转换为具有高度准直性光束的换能器。

[7] Laser machining is localized, non-contact machining and is almost reacting-force free.

句意：激光加工可以实现局部的非接触加工，因此对加工件几乎没有作用力。

[8] Heat affected zone (HAZ) in laser machining is relatively narrow and the re-solidified layer is of micron dimensions.

句意：激光加工中的热影响区相对较窄，其重凝固层厚度只有几微米。

[9] The sonotrode is a piece of metal or tool that is exposed to *ultrasonic* vibration, and then gives this vibratory energy in an element to excite the abrasive grains in the slurry.

句意：音极是暴露在超声波振动中的一小块金属或工具，它将振动能传给某个元件，从而激励浆料中的磨粒。

Glossary

abrade　*v.* 磨蚀；磨损；擦伤
abrasive　*adj.* 研磨的；磨损的　*n.* 研磨剂，研磨料
accomplish　*v.* 完成；实现；成就
acid　*n.* 酸性物质，酸　*adj.* 酸的，酸性的
acronym　*n.* 缩写字；字头语
adhesion　*n.* 附着（力）；粘连（作用）
alkaline solution　碱性溶液
amplifier　*n.* 放大器；扩音机
anode　*n.*（电）阳极，正极
anodic dissolution　阳极溶解
artwork　*n.* 布线图；原图；工艺品
aspect ratio　深宽（径）比
booster　*n.* 加强剂量；增强器；增压器；助力器
boron carbide　碳化硼
burr　*n.* 毛刺；毛边　*v.* 在……上形成毛边
capacitor bank　电容器组，电容器组合
cathode　*n.*（电）阴极，负极
chemical blanking　化学造形（型）
chemical solution　化学溶液
circulate　*v.*（使）循环，（使）流通；（使）传播
clamp　*v.* 夹住；夹紧　*n.* 夹子；钳子
collimate　*v.* 照准，瞄准；使成平行

colloquially *adv.* 用通俗语

concentrate *n.* 集中；浓缩；专心

corrosion *n.* 腐蚀；侵蚀

crack *v.*（使）破裂；发出爆裂声 *n.* 裂缝，裂痕

crystal *n.* 水晶；石英；晶体 *adj.* 水晶（制）的；透明的

debris *n.* 堆积物；洗涤残余物；料屑

deburr *v.* 去毛刺，清理毛刺

deformable *adj.* 可变形的

deplate *v.* 除镀层，退镀

dielectric *n.* 电介质

directivity *n.* 定向性；指向性；方向性

dissolution *n.* 分解；溶解

dominate *v.* 控制；支配；统治

dross *n.* 溶渣；熔渣；浮渣；糟粕

duplicate *v.* 复制，复写；使加倍

electrode *n.* 电极；电焊条

electrolytic *adj.* 电解的，由电解产生的

etch *v.* 蚀刻

etchant *n.* 刻蚀；蚀刻剂

extremely *adv.* 极端地；极其，非常

extensively *adv.* 广大地，广泛地；大规模地

Faraday law 法拉第定律

feasible *adj.* 切实可行的；可能的

flexible *adj.* 柔性的；灵活的

flush *v.* 平接；冲洗 *adj.* 齐平的，同高的；埋入的

focused power density 聚焦功率密度

fragile *adj.* 脆性的，易碎的

grit *n.* 硬渣；磨光粉；棱角粒料；粒度

humidity *n.* 湿度；湿气

immerse *v.*（使）浸没

independent *adj.* 独立的；自主的；单独的

insulate *v.*（使）绝缘；（使）隔离

intricate *adj.* 复杂精细的；错综复杂的

irradiation *n.* 发光；放射；照射

kerf *n.* 切口；截口

maskant *n.* 保护层；掩蔽体

match　*v.* 匹配，相称
microscopic　*adj.* 微观的；显微镜的；高倍放大的
negligible　*adj.* 可忽略的，无关紧要的；可不计的
oscillator　*n.* 振动器；发振器
particle　*n.* 粒子；微粒
penetration　*n.* 穿入；渗透
photosensitive　*adj.* 感光性的
pocket　*n.* 溶蚀坑；凹处
polycrystalline　*adj.* 复晶的，多晶的
ram　*n.* 撞锤，锤体；柱塞；挑杆
reagent　*n.* 试剂（导致化学反应）
recur　*v.* 复发；重现
residual stress　残余应力
satisfactory　*adj.* 良好的；符合要求的；令人满意的
sensitive　*adj.* 敏感的；易受伤害的；灵敏的
servomechanism　*n.* 伺服机构，伺服传动
shear　*v.* 剪切；切变
silicon　*n.* 硅；矽
slender　*adj.* 细长的；微薄的
slurry　*n.* 料液，料将；浆液，浆料
sonotrode　*n.* 音极；超声波发生器
spark eroding　电火花腐蚀
spark machining　电火花加工
spark wire erosion　线切割
spectral purity　光谱纯度
stroke　*n.* 冲程；行程，冲量冲击；冲孔
stock removal rate　材料去除率
sufficiently　*adv.* 足够地，充分地
suitably　*adv.* 合适地；适宜地；适当地
surface texturing　表面织构（化）
suspend　*v.* 悬浮；悬挂；中止
technique　*n.* 技术；工艺方法；技能
thermoelectric　*adj.* 热电的
transducer　*n.* 换能器；变能器
transient　*adj.* 短暂的，瞬时的
trepan　*v.* 环锯　*n.* 钻孔机；凿井机

undercut *n.* 钻蚀，掏蚀；侧凹
vibratory *adj.* 振动的，振动性的
virtually *adv.* 实际地；实质地；事实上
wear *v.* 磨损；耗损

Unit 7　High Speed Cutting (HSC)

7.1　Definition

With increasing demands for higher productivity and lower production costs, investigations have been carried out since the late 1950s to increase the material removal rate in machining, particularly for applications in the aerospace and automotive industries. One obvious possibility is to increase the cutting speed.

As a general guide, an approximate range of cutting speeds may be defined as follows:
1. High speed: 600—1800 m/min (2000—6000 ft/min);
2. Very high speed: 1800—18000 m/min (6000—60000 ft/min);
3. Ultrahigh speed: > 18000 m/min.

High-speed machining (HSM) is characterized by cutting metals and alloys at very high cutting speeds and feeds. For a given work material, the machining is said to be in the high-speed range, if the cutting speed lies between 5 and 10 times of its conventional cutting speed. The most significant difference between conventional machining (CM) and HSM is that localized overheating at primary shear zone may lead to thermal softening of the work material. The resulting thermal softening causes substantial reduction in cutting force and flow stress.

7.2　Introduction to High Speed Cutting

High speed machining was patented by Salomon in the 1930s. Based on metal cutting studies made by Salomon on steel, non-ferrous and light metals at cutting speeds of 440 m/min (steel), 1600 m/min (bronze), 2840 m/min (copper) and up to 6500 m/min (aluminum), the essential result described was the fact that from a certain cutting speed upward machining temperatures start dropping.

These studies have indicated that high-speed machining can be economical for certain applications. Consequently, it is now implemented for the machining of aircraft-turbine components and automotive engines with five to ten times the productivity of traditional machining. High-speed machining of complex 3- and 5-axis contours has been made possible only recently by advances in CNC control technology.[1]

Benefits of HSC:
- Reduced machining time (enhanced machining efficiency)
- Reduced mechanical stresses (reduced distorsion of parts)

- Reduced heating of the part
- Improved surface quality and dimensional accuracy
- Ability to use smaller tools (allow for micro-mechanical machining)

The high speed machining has been widely used in aerospace and mold manufacturing because of its lower cutting cost and good mechanical properties of the cutting tools. Therefore, the mechanical properties of cutting tools become the key factors to ensure the machining efficiency. Ceramic cutting tools have already been widely used for machining hard materials due to their unique mechanical properties, such as silicon nitride (Si_3N_4) ceramic cutting tools, which have become the promising ceramic cutting tools because of their high hardness, strength, and fracture toughness. However, the limitations of their applications are insufficient wear resistance and inferior thermal shock resistance, which will cause high wear rate. Diamond and cubic boron nitride (CBN) as conventional superhard materials have found widespread industrial applications. Diamond exhibits some excellent physical and chemical properties, such as extreme hardness, high melting temperature and thermal conductivity, however, it is prone to oxidation even at moderate temperatures and not suitable for high-speed cutting and polishing of ferrous alloys due to its poor chemical inertness.[2] While CBN, the super abrasive of choice for machining hard ferrous steels, has superior thermal stability and reaction resistance (1376K for CBN and 953K for diamond), however, its hardness is only about half as hard as diamond (47 GPa for CBN and 85 GPa for diamond).

Owing to the high surface qualities it is possible in many cases to eliminate subsequent finish machining entirely or in part. An example to be mentioned is turbine manufacturing where blades are already machined by milling alone and no longer by grinding.

Whereas the HSC technology found its first use in the aerospace industry, present applications come not only from tool- and die-making, but also the production of high-accuracy parts, as well as thin-walled parts (Table 7.1).

Table 7.1 Ranges of HSC applications

Ranges of application	Examples
Aerospace industry	Structural parts (ends)
Composite machining	Turbine blades
Automotive industry	Pattern making Forming dies (sheet metal forming) Injection moulds
Consumer goods-, electrical-and electronic-industry	Electrodes (graphite/copper) Die inserts (hardened) models
Handling technology, energy generation	Compressor wheels, blades, housings

In milling, HSC does not necessarily mean using very high spindle speeds, since, if with milling cutters of greater diameter, HSC can be performed at lower speeds. During finishing of hardened steel materials, when using HSC, we have cutting speed and feed values that are approximately 4 to 6 times greater than the conventional cutting values. HSC is applied more and more to highly productive machining of housings, small-sized and medium-sized components—from roughing to finishing, sometimes even to fine finishing.

7.3 High Speed Cutting Techniques

High speed cutting, with many advantages in productivity and efficiency, currently finds its way into almost every field of machining. Let us compare the kinematics background of turning and milling procedures. In turning, where the kinematic mechanism is based on a rotating workpiece, it is much more difficult to cope with the huge, quickly rotating weights and safe workpiece at high speeds, and thus the conditions for HSC use are considerably less appropriate than in milling in general.[3]

Consequently, the main focus of HSC is on the milling processes. Thus the following considerations are dedicated to milling technology. As mentioned in the introduction, high speed machining will only succeed if there is a perfect interaction among machine tool, tool, workpiece and tool clamping techniques, cutting fluids, cutting parameters, such as spindle speeds, cutting speeds and feeds. High speed machining of aircraft components, such as aluminum formers and ribs, whose cutting sometimes requires up to 95% of the total energy or efforts of the process. The cutting speeds that can be achieved currently range from 1000 to 7000 m/min, and maximal feeds are up to 30 m/min.

High speed milling of steel and castings, especially finishing, is becoming more and more important, because production times are dramatically reduced thanks to the much higher feed rates.

For hardness values from 46 to 63 HRC, high speed milling may replace even the costly cavity sinking technology by choosing suitable milling cutters and selecting appropriate process parameters.

7.4 HSC Machines

High speed machining is only possible if all elements of the system of machine tool-tools-workpiece are optimally matched. The stringent kinematic and dynamic requirements that must be fulfilled by the corresponding machine design for different purposes of use demand modular approaches with innovative solutions in machine building, e.g. granite castings for the frames, and advanced drive- and control technology.[4]

For this reason, the implementation of high speed milling technology in industry resulted in a wide variety of high speed machining centers and machines, as required for different machining tasks, such as light alloy machining in the aerospace industry, with cutting requiring a great expenditure of energy, or finishing of hardened steel dies in the tool- and die-making industry. Some of the significant assemblies and components of a high speed machine system will be discussed briefly.

Spindle rotational speeds today may range up to 40000 rpm, although the automotive industry, for example, generally limits them to 15000 rpm for better reliability and less downtime should a failure occur.

Spindle designs for high speeds generally involve an integral electric motor (motorized spindle). The armature is built onto the shaft and the stator is placed in the wall of the spindle housing. The bearings may be rolling element or hydrostatic; the latter requires less space than the former.

Motorized spindles with ball bearing have been proven effective as main spindles because of their good dynamic performance.

Among the feed drives, the electromechanical servo linear motors are dominant, but the linear direct drives, which enable much higher feed rates (>100 m/min) and acceleration values of 5 to 10 g (50—100 m/s^2) are only at the experimental stage. For small- and medium-sized HSC machines, the machine frames are made of granite, whose damping capacity is 6 to 10 times higher than grey cast iron, and whose thermal expansion coefficient is only 1/3 to 1/5 of steel.

Concerning the axis allocation, the three major axes X, Y and Z are, as a rule, defined as Cartesian linear axes. In addition to these, circular- and swiveling axes are implemented for the transition from 3- to 5-axis milling in different variants.

Notes

[1] These studies have indicated that high-speed machining can be economical for certain applications. Consequently, it is now implemented for the machining of aircraft-turbine components and automotive engines with five to ten times the productivity of traditional machining. High-speed machining of complex 3- and 5-axis contours has been made possible only recently by advances in CNC control technology.

句意：研究表明，高速加工在某些应用领域是很经济划算的。所以，它常常用于加工飞机发动机零部件和汽车发动机，其加工效率是传统加工方式的5～10倍。由于CNC技术的发展，现在复杂零件的3轴和5轴高速切削加工已经可以实现。

[2] Diamond exhibits some excellent physical and chemical properties, such as extreme hardness, high melting temperature and thermal conductivity, however, it is prone to oxidation

even at moderate temperatures and not suitable for high-speed cutting and polishing of ferrous alloys due to its poor chemical inertness.

句意：金刚石呈现了一些优良的物理化学性能，如高硬度、高熔点、良好的导热性等。但是即使在中等温度条件下它也极易氧化，所以由于它的化学惰性不好，不适合对铁合金材料进行高速切削和抛光。

[3] Let us compare the kinematics background of turning and milling procedures. In turning, where the kinematic mechanism is based on a rotating workpiece, it is much more difficult to cope with the huge, quickly rotating weights and safe workpiece chucking at high speeds, and thus the conditions for HSC use are considerably less appropriate than in milling in general.

句意：先让我们比较一下车削和铣削过程的运动学机理。车削是对旋转的工件进行加工。显然，要适应高速旋转的大型重块零件及安全的工件装夹是相当困难的，因此在一般情况下，高速车削的工艺条件不如铣削。

[4] The stringent kinematic and dynamic requirements that must be fulfilled by the corresponding machine design for different purposes of use demand modular approaches with innovative solutions in machine building, e.g., granite castings for the frames, and advanced drive- and control technology.

句意：因此必须根据不同的使用目的来进行相应的机床设计以满足运动学和动力学方面的严格要求，要采用模块化设计手段在机床制造上采用创新方案，如采用花岗岩浇注的床身设计，引入先进的驱动和控制技术。

Glossary

aerospace   *n.* 航空航天空间；*adj.* 航空航天的
aircraft turbine   航空发动机
armature   *n.* 电枢（电机的部件），转子
bench   *n.* 长凳，工作台，试验台
breakdown   *n.* 故障，损坏
carbide   *n.* 碳化物
cubic boron nitride (CBN)   *n.* 立方氮化硼
cavity sinking   （用电火花）加工零件凹型
cemented carbide   硬质合金
chucking   *n.* 夹紧、卡紧
die-making   模具制造
damping capacity   吸震能力

downtime   *n.* （工厂等由于检修，待料等的）停工期，故障停机时间
electromechanical   *adj.* 电动机械的，机电的
expenditure   *n.* 支出，花费
flowing chip   流动的切屑
finishing   *n.* 精整加，精加工，最后的修整
feed   *n.* 进给量，馈送
ferrous material   铁合金材料，钢铁材料
former   *n.* 模型，样板
fracture   *n.* 破裂，断裂
housing   *n.* 外壳，机器等的防护外壳或外罩
hydrostatic   *adj.* 流体静力学的，静水力学的
modular   *adj.* 模块化的，有标准组件的，组装式的
overheating   *n.* 过热
primary shear zone   主剪切区
rib   *n.* 筋板，肋骨
roughing   *n.* 初步加工，粗加工
stator   *n.* 定子
silicon nitride   氮化硅
superhard   *adj.* 超硬的
servo linear motors   伺服直线电机
swiveling swiveling   *adj.* 旋转的，转动的
toughness   *n.* [力] 韧性，坚韧
thermal shock resistance   抗热冲击性

Unit 8　Tolerances and Fits

8.1　Introduction

Quality and accuracy are major considerations in making machine parts or structures. Interchangeable parts require a high degree of accuracy to fit together. With increasing accuracy or less variation in the dimension, the labor and machinery required to manufacture a part is more cost intensive.[1] Any manufacturer should have a thorough knowledge of the tolerances to increase the quality and reliability of a manufactured part with the least expense.

An engineering drawing must be properly dimensioned in order to convey the designer's intent to the end user. Dimensions of parts given on blueprints and manufactured to those dimensions should be exactly alike and fit properly. Unfortunately, it is impossible to make things to an exact dimension. Most dimensions have a varying degree of accuracy and a means of specifying acceptable limitations in dimensional variance so that a manufactured part will be accepted and still function. It is necessary that the dimensions, shapes and mutual position of surfaces of individual parts be kept within a certain accuracy to achieve their correct and reliable functioning. Routine production processes do not allow maintenance (or measurement) of the given geometrical properties with absolute accuracy.[2] Actual surfaces of the produced parts therefore differ from ideal surfaces prescribed in drawings. Deviations of actual surfaces are divided into four groups to enable assessment, prescription and checking of the permitted inaccuracy during production:

1. Dimensional deviations;
2. Shape deviations;
3. Position deviations;
4. Surface roughness deviations.

As mentioned above, it is principally impossible to produce machine parts with absolute dimensional accuracy. In fact, it is not necessary or useful. It is quite sufficient that the actual dimension of the part is found between two limit dimensions and a permissible deviation is kept with production to ensure correct functioning of engineering products. The required level of accuracy of the given part is then given by the dimensional tolerance which is prescribed in the drawing. The production accuracy is prescribed with regards to the functionality of the product and to the economy of production as well. The principal factor used to set a tolerance for a dimension should be the function of the feature being controlled by the dimension. Unnecessarily tight tolerances lead to high cost of manufacture resulting from more expensive manufacturing methods and higher reject rates.

8.2 Tolerances

Tolerance is the total amount that a specific dimension is permitted to vary. It is the difference between the maximum and the minimum limits for the dimension. To understand tolerances, we should understand some of the definitions associated with the determination of a tolerance. These definitions may be generally categorized as relating to size, allowance, or fit.

Size. The size of an object or its mate is known as nominal, basic, or design size.

Allowance. The intentional difference between the maximum material limits of mating parts. It is the difference between the largest allowable shaft size and the smallest hole diameter.[3] This is a minimum clearance (positive allowance) or maximum interference (negative allowance) between mating parts. The quality of the fit is characterized by the allowance value.

Fit. Clearance, interference, or transition fit refer to how the object fits an assembly.

To specify the size of an object, we dimension it with a nominal size, limit size, and actual size shown in Fig.8.1.

Nominal size (basic size). It is the designation used for general identification and usually expressed in common fractions. It is an exact theoretical size of a part from which limit dimensions are computed. The nominal size of a part should be selected from the preferred dimension series indicated by the national standard.[4] [shown in Fig.8.1(a)]

Limit size. It includes upper limit size and lower limit size which are allowable extreme sizes of a part. Fox example, the limits for a hole are 1.500 (lower limit) and 1.501(upper limit) and for a shaft 1.499 (upper limit) and 1.497 (lower limit), shown in Fig.8.1(b).

Actual size. It is the measured size of the finished part, shown in Fig.8.1(c).

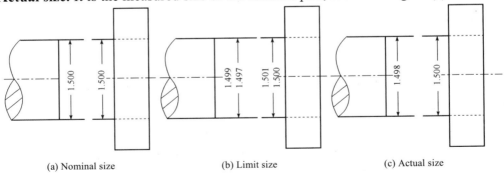

(a) Nominal size   (b) Limit size   (c) Actual size

Fig.8.1 Size designation

The units in the figures in this book are millimeters.

Deviations. The upper and lower deviations obtained by subtracting basic size from limit size. For example, a hole dimensioned as 2.000±0.0004, thus upper deviation is +0.0004 and lower deviation is −0.0004.

Tolerances can be expressed in either of two ways.

Bilateral tolerances. A bilateral tolerance is a tolerance in which variation is permitted in both directions from a specified dimension. Example is 2.000±0.0004. For this expression, the dimension of the part would be permitted to vary between 1.996 and 2.004 for a total tolerance of 0.008.

The actual size of the object may be larger or smaller than the stated size limitation if there can be equal variation in both directions. The plus and minus limitations combine to form a single value.[5]

Unilateral tolerances. A unilateral tolerance is a tolerance in which variation is permitted only in one direction from the specified dimension. Example is $2.000^{+0.008}_{+0.000}$. For this expression, the dimension of the part would be permitted to vary between 2.000 and 2.008 for a total tolerance of 0.008.

Unilateral tolerance expression has the advantage that they are easier to check on drawings and that the change in the tolerance can be made with the least disturbance to other dimensions.

Fig.8.2 indicates tolerance.

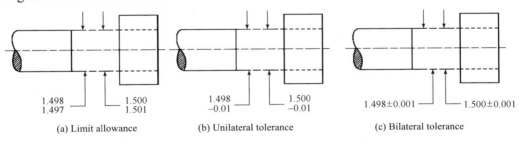

(a) Limit allowance   (b) Unilateral tolerance   (c) Bilateral tolerance

Fig.8.2　Tolerance annotation

8.3　Fits

How mating parts or assemblies fit together with component parts is referred to as fit, which includes clearance fit, interference fit, or transition fit, see Fig.8.3.

Fit is the general range of tightness resulting from the application of a specific combination of allowance and tolerances in the design of mating parts.

Clearance fit. It is a fit enabling a clearance between the hole and shaft in the coupling. The lower limit size of the hole is greater or at least equal to the upper limit size of the shaft.

Interference fit. It is a fit always ensuring some interference between the hole and shaft in the coupling. The upper limit size of the hole is smaller or at least equal to the lower limit size of the shaft.

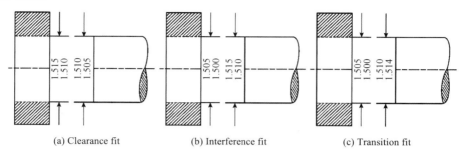

Fig.8.3   Type of fits

Transition fit. It is a fit where (depending on the actual sizes of the hole and shaft) both clearance and interference may occur in the coupling. Tolerance zones of the hole and shaft partly or completely interfere.

8.4   ISO System of Limits and Fits

The standard ISO is used as an international standard for linear dimension tolerances and has been accepted in most industrially developed countries as a national standard. The ISO system of tolerances and fits can be applied in tolerances and deviations of smooth parts and for fits created by their coupling. It is used particularly for cylindrical parts with round sections. Tolerances and deviations in this standard can also be applied in smooth parts of other sections. Similarly, the system can be used for coupling (fits) of cylindrical parts and fits with parts having two parallel surfaces (e.g., fits of keys in grooves). The term "shaft", used in this standard has a wide meaning and serves for specification of all outer elements of the part, including those elements which do not have cylindrical shapes. Also, the term "hole" can be used for specification of all inner elements regardless of their shape.

In ISO system, each part has a basic size whose limit dimensions are specified using the upper and lower deviations.[6] In case of a fit, the basic size of both connected elements must be the same.

The tolerance of a basic size is defined as the difference between the upper and lower limit dimensions of the part. In order to meet the requirements of various production branches for accuracy of the product, the system ISO implements 20 grades of accuracy. Each of the tolerances of this system is marked "IT" with attached grade of accuracy (IT01, IT0, IT1,···, IT18).

The tolerance zone is defined as an annular zone limited by the upper and lower limit dimensions of the part. The tolerance zone is therefore determined by the amount of the tolerance and its position related to the basic size. The position of the tolerance zone, related to the basic size (zero line), is determined in the ISO system by a so-called basic deviation. The system ISO defines 28 classes of basic deviations for holes. These classes are marked by capital letters (A, B, C,···, ZC). The tolerance zone for the specified dimensions is prescribed in the drawing by a tolerance mark, which consists of a letter marking of the basic deviation and a numerical marking of the tolerance grade (e.g., H7, H8, D5, etc.). Though the general sets of basic deviations (A,···, ZC) and tolerance grades (IT1,···, IT18) can be used for prescriptions of hole tolerance zones by their mutual combinations, in practice only a limited range of tolerance zones is used. The system ISO also defines 28 classes of basic deviations for shafts. These classes are marked by lower case letters (a, b, c,···, zc). The tolerance zone for the specified dimensions is prescribed in the drawing by a tolerance mark, which consists of a letter marking of the basic deviation and a numerical marking of the tolerance grade (e.g. h7, h6, g5, etc.). Though the general sets of basic deviations (a,···, zc) and tolerance grades (IT1,···, IT18) can be used for prescriptions of shaft tolerance zones by their mutual combinations, in practice only a limited range of tolerance zones is used.

Although there can be generally coupled parts without any tolerance zones, only two methods of coupling of holes and shafts, namely basic hole system and basic shaft system, are recommended due to constructional, technological and economic reasons.[7]

The basic hole system is a system of fits in which the design of the hole is the system basic size and the allowance applies to the shaft. When specifying the tolerances for a hole and cylinder and determining their dimensions, we should begin calculating by assuming either the minimum (smallest) hole or the maximum (largest) shaft size if they are to fit together well. The desired clearances and interferences in the fit are achieved by combinations of various shaft tolerance zones with the hole tolerance zone "H". In this system of tolerances and fits, the lower deviation of the hole is always equal to zero. Fig.8.4 illustrates the basic hole system. In the illustration, the minimum hole size (1.500) is the basic size. To calculate the maximum diameter of the shaft, assume an allowance of 0.003 and subtract that from the basic hole size. Arbitrarily selecting a tolerance of 0.002 for both the hole and the shaft gives a maximum hole (1.502) and minimum shaft (1.495). The minimum clearance fit is the difference between the smallest hole (1.500) and the largest shaft (1.497) or 0.003. The maximum clearance fit is the difference between the largest hole (1.502) and the smallest shaft (1.495) or 0.007. By adding the allowance (0.003) to the basic hole size (1.500), the maximum shaft size of an interference fit is obtained. To convert basic hole size to basic shaft size, we subtract the allowance for a clearance fit or add it for an interference fit.

The basic shaft system is a system of fits in which the design size of the shaft system is the basic size and the allowance applies to the hole. The desired clearances and interferences in the fit are achieved by combinations of various hole tolerance zones with the shaft tolerance zone "h". In this system of tolerances and fits, the upper deviation of the shaft is always equal to zero. Fig.8.5 illustrates the basic shaft system. In the illustration, the maximum shaft size is the basic size. To obtain the minimum hole diameter, assume an allowance of 0.003 and add it to the basic shaft size. Arbitrarily selecting a tolerance of 0.002, add the tolerance to the hole and shaft can obtain the maximum hole (1.505) and the minimum shaft (1.498). The minimum clearance (0.003) is the difference between the smallest hole (1.503) and the largest shaft (1.500). The maximum clearance (0.007) is the difference between the largest hole (1.505) and the smallest shaft (1.498).

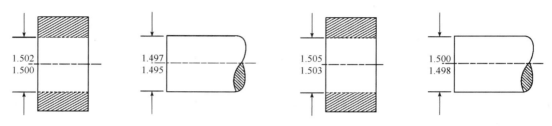

Fig.8.4　The basic hole system　　　　　　Fig.8.5　The basic shaft system

Any manufacturer should have a thorough knowledge of the tolerances and fits to increase the quality and reliability of a manufactured part with the least expense. The increased compatibility of a certain product with mating parts made by different manufacturers makes it more viable in market competition. Moreover, a knowledge of tolerances and fits is essential for manufacturing with least material wastage, and hence good tolerancing principles followed by manufacturers are essential in reaping profits.

Notes

[1] With increasing accuracy or less variation in the dimension, the labor and machinery required to manufacture a part is more cost intensive.

句意：随着精度的增加和尺寸变化的减小，制造零件所需的劳动力和机器更趋于成本密集型。

[2] Routine production processes do not allow maintenance (or measurement) of the given geometrical properties with absolute accuracy.

句意：常规的生产过程中不可以对具有绝对精度的给定几何特征进行维修或测量。

[3] It is the difference between the largest allowable shaft size and the smallest hole diameter

句意：它是所允许的最大轴径与最小孔径之间的差值。

[4] The nominal size of a part should be selected from the preferred dimension series indicated by the national standard.

句意：零件的标称尺寸应从国家标准规定的优先尺寸中选取。

[5] The plus and minus limitations combine to form a single value.

句意：正、负极限组合起来就构成一个单一值。

[6] In ISO system, each part has a basic size whose limit dimensions are specified using the upper and lower deviations.

句意：在 ISO 体系中，每一个零件都有一个基本尺寸，其极限尺寸是采用上、下偏差来定义的。

[7] Although there can be generally coupled parts without any tolerance zones, only two methods of coupling of holes and shafts, namely basic hole system and basic shaft system, are recommended due to constructional, technological and economic reasons.

句意：尽管存在没有公差带的一般装配零件，但基于结构、技术和经济原因，这里只推荐两种孔和轴的连接方法，即基孔制和基轴制。

Glossary

acceptable　*adj.* 可接受的；合意的
accuracy　*n.* 精度；准确度；精密
achieve　*v.* 完成，实现；达到，得到
actual　*adj.* 实际的；现行的
arbitrarily　*adv.* 任意地
assessment　*n.* 评定，估计，评估
basic deviation　基本偏差
basic hole system　基孔制
basic shaft system　基轴制
bilateral tolerance　双向（边）公差
blueprint　*n.* 蓝图；计划
branch　*n.* 分支；分科；分部
capital letter　大写字母
clearance fit　间隙配合
combination　*n.* 联合；合并；结合
common fraction　简分数

convey　*v.* 传达，传递
cylindrical　*adj.* 圆柱的
definition　*n.* 定义；释义；定界
designation　*n.* 指定；名称；命名；符号名称
deviation　*n.* 偏离；背离；偏差数
dimension　*n.* 尺寸，尺度；维数；容积，面积
disturbance　*n.* 扰乱；扰动；干扰
drawing　*n.* 图样；绘图
exact　*adj.* 精确的，准确的；确切的
fit　*n.* 配合；适合　*v.* 使适合；安装；符合　*adj.* 适合的
functionality　*n.* 功能性；泛函性
geometrical　*adj.* 几何的，几何学的
identification　*n.* 辨识；识别；鉴别；确认
illustrate　*v.* 举例说明；阐明；图解
industrially　*adv.* 工业上地，产业上地
interchangeable　*adj.* 可互换的
interference fit　过盈配合
intent　*n.* 意图，目的，意向
least　*adj.* 最小的；最少的　*adv.* 最少地；最小地
limitation　*n.* 极限；限度；局限
lower case letter　小写字母
maintenance　*n.* 维护；维持；保养，维修
maximum　*n.* 最大量；极大；最大限度　*adj.* 最大量的，极大的，最大限度的
minimum　*n.* 最小量　*adj.* 最小的
mutual　*adj.* 相互的；共同的
national standard　国家标准
nominal size　标称尺寸
permissible　*adj.* 允许的，准许的
prescribe　*v.* 规定；指示，命令
prescription　*n.* 规定；指示
principal　*adj.* 重要的，主要的
reap　*v.* 收获；获得
recommend　*v.* 推荐；介绍；建议
reject rate　废品率
reliability　*n.* 可靠性
roughness　*n.* 粗糙度；粗糙率；不平整度

routine　*adj.* 常规的、例行的　*n.* 常规，例行公事
shaft　*n.* 轴；杆状物
specification　*n.* 详述；说明；规格；说明书
subtract　*v.* 减，减去；去掉
sufficient　*adj.* 足够的，充分的
theoretical size　理论尺寸
thorough　*adj.* 完全的，全面的；彻底的
tight tolerance　紧公差，小公差
tightness　*n.* 致密度；紧度；密封度；坚固；紧密
tolerance　*n.* 公差
tolerance grade　公差等级
tolerance zone　公差带
transition fit　过渡配合
unilateral tolerance　单向（边）公差
variance　*n.* 变化；变量；不同
variation　*n.* 变化，改变；变种，变异；变化量
zero line　零线；基准线

Unit 9   Location and Fixtures

9.1   Introduction

To ensure that the workpiece is produced according to the specified shape, dimensions and tolerances, it is essential that workpiece should be appropriately located and clamped on the machine tool.[1] Production devices are generally workholders with/without tool guiding/setting arrangement. These are called jigs and fixtures. A fixture is a production tool that locates, holds and supports the workpiece securely so that each part is machined within the specified limits. It must correctly locate a workpiece in a given orientation with respect to a cutting tool or measuring device, or with respect to another component, as for instance in assembly. Such location must be invariant in the sense that the devices must clamp and secure the workpiece in that location for the particular processing operation. Fixtures are generally used in machining (milling, planing, shaping and turning, etc.) and other manufacturing operations, such as welding, heat treatment, machining inspection and assembly, etc. Jigs are provided with tool guiding elements such as drill bushes. They direct the tool to the correct position on the workpiece. Jigs are rarely clamped on the machine table because it is necessary to move the jig on the table to align the various bushing in the jig with the machine spindle. Jigs may be divided into two general classes, boring jigs and drill jigs.

9.2   Advantages of Jigs and Fixtures

Productivity. Jigs and fixtures eliminate individual marking, positioning and frequent checking, which reduces operation time and increase productivity.

Interchangeability. Jigs and fixtures facilitate uniform quality in manufacture. There is no need for selective assembly. Any part of machine fits properly in assembly and all similar components are interchangeable.

Skill reduction. Jigs and fixtures simplify locating and clamping of the workpieces. Tool guiding elements ensure correct position of the tools with respect to the workpieces. There is no need for skillful setting of the workpiece and tool. Any average person can be trained to use jigs and fixtures, the replacement of a skilled workman with unskilled labor can effect substantial saving in labor cost.

Cost reduction. Higher production rate, reduction in scrap, easy assembly and savings in labor costs result in substantial reduction in the cost of workpieces produced with jigs and fixtures.

9.3 Location of Workpiece

A body that is completely free in space has twelve degrees of freedom shown in Fig.9.1. It can rotate about or have linear movement along each of the three mutually perpendicular axes *XX*, *YY* or *ZZ*, so it has six freedoms of rotation and six freedoms of translation.[2] When workpiece is located, it must be constrained from moving in any direction, so these freedoms are eliminated or restricted to ensure that the operation is performed with the required accuracy. This can be done by six locations in the case of the body shown in Fig.9.1, which is known as six-point locating principle. It is illustrated in Fig.9.2.

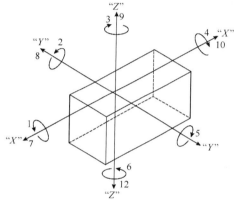
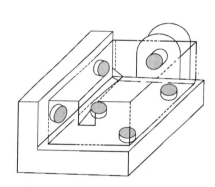

Fig.9.1 The twelve degrees of freedom   Fig.9.2 Six-point location

The base of the component is resting on three locating pins which is the minimum number of points upon which it will firmly seat. The three adjacent locating surfaces of the workpiece are resting against 3, 2 and 1 pins respectively, which prevent nine degrees of freedom. The rest three degrees of freedom are arrested by three external forces usually provided directly by clamping. Some of such forces may be attained by friction. If more than six points are used, the additional points will be surplus and unnecessary and would therefore be redundant constraints.

There are three general forms of location: plane, concentric, and radial.[3] Plane locators locate a workpiece from any surface. The surface may be flat, curved, or have an irregular contour. In most applications, plane-locating devices locate a part by its external surfaces, see Fig.9.3(a). Concentric locators, for the most part, locate a workpiece from a central axis. This axis may or may not be in the center of the workpiece. The most-common type of concentric location is a locating pin placed in a hole. Some workpieces, however, might have a cylindrical projection that requires a locating hole in the fixture, as shown in Fig.9.3(b). The third type of location is radial. Radial locators restrict the movement of a workpiece around a concentric locator, Fig.9.3(c). In many cases, locating is performed by a combination of the three locating methods.

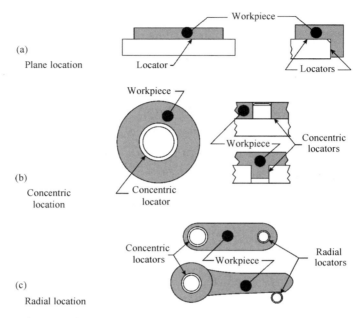

Fig.9.3　The three forms of location: plane, concentric, and radial

In machining practice, depending on concrete requirements, not all the twelve degrees of freedom are necessary to be restricted.[4] In Fig.9.4(a), on the workpiece, a slot should be machined and so its freedom of movement in all directions should be restricted. In Fig.9.4(b), however, the machining of a stepped surface needs only ten degrees of freedom being restricted. And from Fig.9.4(c), it can be seen that since only the upper surface should be machined (e.g., 2 surface grinding), restricting three degrees of freedom is enough.

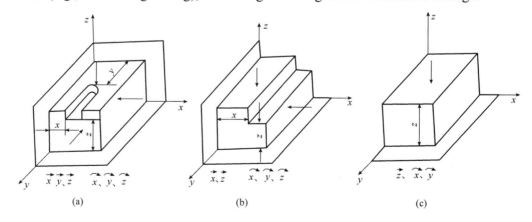

Fig.9.4　DOF restriction on different machining requirements

Another condition to avoid in workholder design is redundant, or duplicate location. Redundant locators restrict the same degree of freedom more than once. The workpieces in Fig.9.5 show several examples. The part at Fig.9.5(a) shows how a flat surface can be redundantly located. The part should be located on only one, not both, side surfaces. Since the sizes of parts can vary, within their tolerances. The example at Fig.9.5(b) points out the same problem with concentric diameters. Either diameter can locate the part, but not both. The example at Fig.9.5(c) shows the difficulty with combining hole and surface location. Either locating method, locating from the holes or from the edges, works well if used alone. When the methods are used together, however, they cause a duplicate condition. The condition may result in parts that cannot be loaded or unloaded as intended.

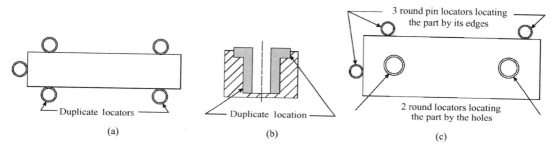

Fig.9.5　Examples of redundant location

9.4　Clamping of Workpiece

As with location, clamping will influence the accuracy and quality of the workpiece and will further influence substantially the speed and efficiency of the operation being carried out on the workpiece. The process of clamping induces a locking effect which, through frictional or some other forms of mechanism, provides a stability of location which cannot be changed until and unless external loading is able to overcome the locking effect. Hence, when a cutting force is producing a load or moment on the workpiece, it is necessary that a sufficient clamping force must be exerted to withstand such actions. Fixtures are used to restrict the freedoms of the workpiece to be machined and locate the workpiece. Clamping device, however, is used to ensure that the fixture, workpiece and location features will not be distorted or damaged under the action of cutting force, gravity, inertial force and centrifugal force, etc.[5] This means that clamping forces should not be excessive but sufficient to hold the workpiece rigidly and should be applied at the points where the workpiece has the support from the solid metal of the fixture body. Clamping forces should be directed towards supporting and locating elements on overhanging or thin sections of the workpiece. In addition, the force should be transmitted to the rigid sections of the fixture body frame.

Cylindrical workpieces located in V-blocks can be clamped using another V-block or clamped in a 3-jaw chuck. The latter is usually more common, especially in turning operations.

9.5 Classes of Fixtures

Fixtures can be classified in terms of applications and features as follows:
1. General purpose fixtures;
2. Special purpose fixtures;
3. Adaptable fixtures;
4. Modular fixtures;
5. Pallet (Fixture movable with workpiece).[6]

Fixtures are also normally classified by the type of machine on which they are used. Fixtures can also be identified by a sub-classification. A fixture is named in accordance with the machine on which it is used. A milling fixture, which is clamped to the table of the milling machine, must be sturdy since vibrations could cause a poorly machined finish. Grinding fixtures may be either clamped in place or held on the table magnetically. A lathe fixture is clamped to the rotating head of the lathe and thus spins while the tool remains stationary. Boring fixtures are used to produce a finer finish and closer tolerance on the diameter of the hole than using a drill jig. A welding fixture is used for holding parts while they are being fused together. Gauging fixtures are used for checking the accuracy of parts after they have been machined and prior to their assembly with other parts.

In spite of that diversity of fixtures, they all have common features with respect to construction and operation principle.

The main components of a fixture include locators, clamping device, guiding and tool-setting elements, fixture body, connecting elements and other components and devices.

Locator. A locator is usually a fixed component of a fixture. It is used to establish and maintain the position of a part in the fixture by constraining the movement of the part through directly contacting or fitting with the locating surface on the workpiece. For workpieces of greater variability in shapes and surface conditions, a locator can also be adjustable.

Clamping device. A clamp is a force-actuating mechanism of a fixture. The forces exerted by the clamps hold a workpiece securely in the fixture against all other external forces to prevent the position of workpiece from changing during the machining operation.

Guiding and tool-setting elements. It is used to specify the position of cutting tool relative to the fixture, such as jig bush.

Fixture body. Fixture body, or tool body, is the major structural element of a fixture. It maintains the spatial relationship between the fixturing elements mentioned above, viz, locators, clamps, supports, and the machine tool on which the part is to be processed.

Connecting elements. Connecting elements are used to specify the position of a fixture relative to machine tool and to fix the fixture on the machine tool, such as orienting keys, T-bolts.

Other components or devices. Other components or devices are used to load and unload workpiece or meet others needs during machining. Such as auxiliary support, loading/unloading device, and indexing device etc.

In order to accurately locate a workpiece in a fixture, movements must be restricted, which is done with locators and clamps. The fixture for the part in Fig.9.4 illustrates this principle of restricting movement. By placing the part on a three-pin base, (three points determine a plane), five directions of movement (2, 5, 1, 4 and 12 shown in Fig.9.1) are restricted, see Fig.9.6. Using pin or button type locators minimizes the chance of error by limiting the area of contact and raising the part above chips.

To restrict the movement of the part around the "ZZ" axis and in direction 8, two more pin-type locators are used, see Fig.9.7. To restrict direction 7, a single-pin locator is used, see Fig.9.8.

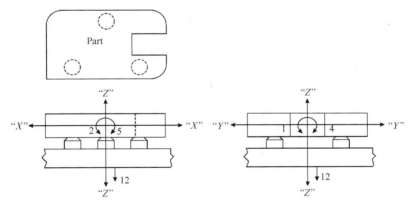

Fig.9.6 Three-pin base restricts five directions of movement

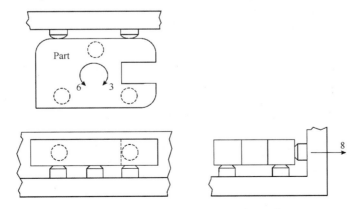

Fig.9.7 Five-pin base restricts eight directions of movement

The remaining movements in directions 9, 10 and 11 are restricted by using a clamping device. This 3-2-1, or 6-point locating principle is most commonly used for square or rectangular parts.

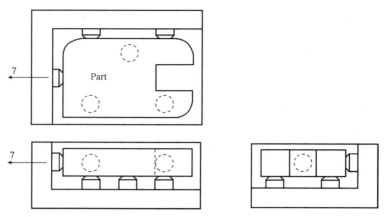

Fig.9.8 Six-pin base restricts nine directions of movement

Notes

[1] To ensure that the workpiece is produced according to the specified shape, dimensions and tolerances, it is essential that workpiece should be appropriately located and clamped on the machine tool.

句意：为了确保按照规定的形状、尺寸和公差将工件加工出来，至关重要的一点就是要将工件恰当地定位并夹紧在机床上。

[2] It can rotate about or have linear movement along each of three mutually perpendicular axes *XX*, *YY* or *ZZ*, so it has six freedoms of rotation and six freedoms of translation. (Here the number of DOF, i.e. degree of freedom, of a free body in space is twelve, not six, because in each axis of *XX*, *YY* and *ZZ*, there are two rotating directions and two translation directions.)

句意：它可以沿相互垂直的 *XX*、*YY* 和 *ZZ* 三个轴进行回转或直线运动，因此它具有 6 个回转自由度和 6 个平移自由度。（空间中的自由物体的自由度应该是 12 而不是 6，因为沿 *XX*、*YY* 和 *ZZ* 各轴都有两个回转方向和平移方向。）①

[3] There are three general forms of location: plane, concentric, and radial.

句意：常见的三种定位形式有：平面定位、同心定位（圆孔定位）和径向定位（外

---

① 在我国高校的教科书上一般都将空间自由刚体的自由度定为 6 个，并且不考虑运动轴的正、负两个方向。——译者注

圆柱面定位）。

[4] In machining practice, depending on concrete requirements, not all the twelve degrees of freedom are necessary to be restricted.

句意：在实际加工过程中，并不是所有 12 个自由度都必须受到限制，要根据具体要求来定。

[5] Fixtures are used to restrict the freedoms of the workpiece to be machined and locate the workpiece, clamping device, however, is used to ensure that the fixture, workpiece and location features will not be distorted or damaged under the action of cutting force, gravity, inertial force and centrifugal force etc.

句意：夹具是用于限制加工件的自由度，并将其定位；而夹紧装置是用于确保夹具、工件和定位特征在切削力、重力、惯性力和离心力等作用下不产生变形和损坏。

[6] Fixtures can be classified in terms of application and features as follows:
1. General purpose fixtures; 2. Special purpose fixtures; 3. Adaptable fixtures; 4. Modular fixtures; 5. Pallet (fixture movable with workpiece).

句意：夹具根据应用和特点可以分为：1. 通用夹具；2. 专用夹具；3. 可调夹具；4. 组合夹具；5. 随行夹具。

Glossary

3-jaw chuck　三爪卡盘
additional　*adj.* 额外的，另外的；附加的
adjacent　*adj.* 相邻的，邻近的；（时间上）紧接着的
align　*v.* 使成一线；对齐；对准
appropriately　*adv.* 恰当地，适当地
attain　*v.* 达到，获得；完成；取得
bush　*n.* 衬套；轴瓦
bushing　*n.* 轴套；套管；衬套
central axis　中心轴
chip　*n.* 薄片；基片；芯片
clamp　*v.* 夹紧，夹住　*n.* 夹子，钳子
clamping device　夹紧元件
concentric　*adj.* 同心的；集中性的
connecting elements　连接元件
constrain　*v.* 约束；强制
contour　*n.* 轮廓

cylindrical *adj.* 圆柱（状）的

dimension *n.* 尺寸，尺度；维（数）

drill jig   钻模

duplicate location   重复定位

eliminate *v.* 消除；排除

essential *adj.* 重要的，主要的；必不可少的，必要的；本质的，实质的

excessive *adj.* 过多的；极度的

exert *v.* 尽（力）；发挥；施加；产生

external *adj.* 外部的，表面的

facilitate *v.* 使容易，使便利；有助于

fixture *n.* 夹具

fixture body   夹具体

frictional *adj.* 摩擦的，摩擦力的

guiding and tool-setting elements   导向元件

influence *v.* 影响，改变 *n.* 影响（力）

interchangeability *n.* 可互换性，可交换性

interchangeable *adj.* 可互换的

invariant *adj.* 不变的，固定的 *n.* 不变量

irregular *adj.* 不规则的，无规律的

jig *n.* 夹紧装置，钻模

limit *n.* 极限；界限；范围 *v.* 限制；限定

locate *v.* 定位；把……置于

locator *n.* 定位元件，定位销

magnetically *adv.* 有磁力地

moment *n.* 力矩；动量

mutually *adv.* 相互地

orientation *n.* 定位；方向，方位

orienting key   定位键，定向键

particular *adj.* 特殊的，特别的，特定的

perpendicular *adj.* 垂直的，正交的 *n.* 垂（直）线；垂直度

pin *n.* 钉，销，栓 *v.* 钉住，别住

productivity *n.* 生产率；生产力

projection *n.* 投影；投射；凸出物

radial *adj.* 半径的，径向的；射线的

redundant constraint   冗余约束

replacement *n.* 取代，替换；代替物，替换物

rotate *v.* （使）旋转，（使）回转
scarp *n.* 废品；废料
securely *adv.* 牢固地；安全地
selective assembly 选择装配
slot *n.* 长孔；狭长的槽 *v.* 开槽于……；在……上开狭长的孔
stepped surface 阶梯面
sturdy *adj.* 坚固的；结实的
substantially *adv.* 本质地，实质地；重大地；相当大地
surplus *adj.* 过剩的，多余的 *n.* 剩余物
tolerance *n.* 公差
translation *n.* 平移，平动
uniform *n.* 均匀；一致 *adj.* 均匀的，一致的
variability *n.* 易变；变化性；变异性
V-block V形块
vibration *n.* 振动；摆动；颤动
workholder *n.* 工件夹具 *v.* 限制；限定

Unit 10　Gear Transmission

10.1　Introduction

A gear drive system is one where a motor turns a series of gears to do work. It plays a very important role of transmitting power, achieving conveying and adjusting torque in numerous industrial equipment like machine tool, automobile, engineering machinery, agricultural machinery and construction machinery.

In general, gear drive is useful for power transmission between two shafts which are near to each other (at most at 1m distance) and also it has maximum efficiency. The gear that supplies the power is called the input gear and the gear that does the actual work at the other end of the gear drive is the output gear. While transmitting power, it is durable compared with other drive systems such as belts, chain drives, etc.

Gears are compact, positive-engagement power transmission elements in gearing transmission; they determine the speed, torque, and direction of rotation of driven machine elements.[1] Gear types may be grouped into five main categories: Spur, Helical, Bevel, Hypoid, and Worm. Typically, shaft orientation, efficiency, and speed determine which of these types should be used for a particular application. This section on gearing and gear drives describes the major gear types; it evaluates how the various gear types are combined into gear drives and considers the principal factors that affect gear drive.

10.2　Spur Gears

Spur gears have straight teeth cut parallel to the rotational axis. The tooth form is based on the involute curve shown in Fig.10.1. Practice has shown that this design accommodates mostly rolling rather than sliding contact of the tooth surfaces.[2] The involute curve is generated during gear machining processes using gear cutters with straight sides.

Near the root of the tooth, however, the tool traces a trochoidal path, Fig.10.2, providing a heavier, and stronger, root section. Because of this geometry, contact between the teeth occurs mostly as rolling rather than sliding. Since less heat is produced by this rolling action, mechanical efficiency of spur gears is high, often up to 99%. Some sliding does occur, however. And because contact is simultaneously across the entire width of the meshing teeth, a continuous series of shocks is produced by the gear. These rapid shocks result in some objectionable operating noise and vibration.[3] Moreover, tooth wear results from shock loads at high speeds. Noise and wear can be minimized with proper lubrication. Spur gears are the least expensive to manufacture and the most commonly used, especially for drives with parallel shafts. The three main classes of spur gears are: external tooth, internal tooth, and rack-and-pinion.

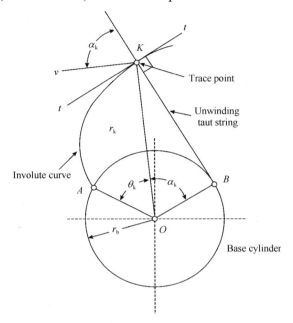

Fig.10.1 Involute generated by unwrapping a cord from a circle

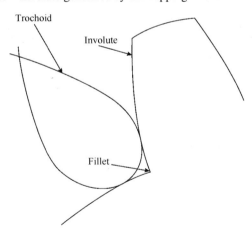

Fig.10.2 Root fillet trochoid generated by straight tooth

External-tooth gears. The most common type of spur gear, Fig.10.3, has teeth cut on the outside perimeter of mating cylindrical wheels, with the larger wheel called the gear and the smaller wheel the pinion. The simplest arrangement of spur gears is a single pair of gears called a single reduction stage, where output rotation is in a direction opposite that of the input. In other words, one is clockwise while the other is counter-clockwise. Higher net reduction is produced with multiple stages in which the driven gear is rigidly connected to a third gear. This third gear then drives a mating fourth gear that serves as output for the second stage. In this manner, several output speeds on different shafts can be produced from a single input rotation.

Internal (ring) gears. Ring gears produce an output rotation that is in the same direction as the input, Fig.10.4. As the name implies, teeth are cut on the inside surface of a cylindrical ring, inside of which are mounted a single external-tooth spur gear or set of external-tooth spur gears, typically consisting of three or four larger spur gears (planets) usually surrounding a smaller central pinion (sun). Normally, the ring gear is stationary, causing the planets to orbit the sun in the same rotational direction as that of the sun. For this reason, this class of gear is often referred to as a planetary system. The orbiting motion of the planets is transmitted to the output shaft by a planet carrier. In an alternative planetary arrangement, the planets may be restrained to orbit the sun and the ring left free to move. This causes the ring gear to rotate in a direction opposite that of the sun. By allowing both the planet carrier and the ring gear to rotate, a differential gear drive is produced, the output speed of one shaft being dependent on the other.

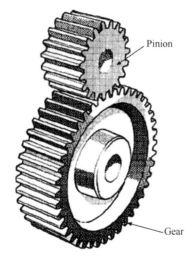

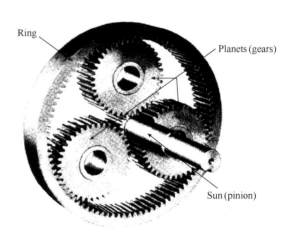

Fig.10.3   External gears   Fig.10.4   Internal gears

Rack-and-pinion gears. A straight bar with teeth cut straight across it, Fig.10.5, is called a rack. Basically, this rack is considered to be a spur gear unrolled and laid out flat. Thus, the rack-and-pinion is a special case of spur gearing. The rack-and-pinion is useful in converting rotary motion to linear and vice versa. Rotation of the pinion produces linear travel of the rack. Conversely, movement of the rack causes the pinion to rotate. The rack-and-pinion is used extensively in machine tools, lift trucks, power shovels, and other heavy machinery where rotary motion of the pinion drives the straight-line action of a reciprocating part. Generally, the rack is operated without a sealed enclosure in these applications, but some type of cover may be provided to keep dirt and other contaminants from accumulating on the working surfaces.[4]

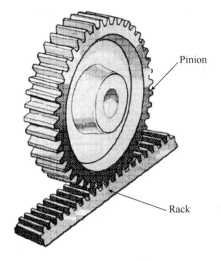

Fig.10.5　Rack-and-pinion gears

10.3　Helical Gears

Helical gearing differs from spur gearing in that helical teeth are cut across the gear face at an angle rather than straight, Fig.10.6. Thus, the contact line of the meshing teeth progresses across the face from the tip at one end to the root of the other, reducing the noise and vibration characteristic of spur gears. Also, several teeth are in contact at any one time, producing a more gradual loading of the teeth that reduces wear substantially. The increased amount of sliding action between helical gear teeth, however, places greater demands on the lubricant to prevent metal-to-metal contact, resulting premature gear failure. Also, since the teeth mesh at an angle, a side thrust load is produced along each gear shaft. Thus, thrust bearings must be used to absorb this load so that the gears are held in proper alignment.

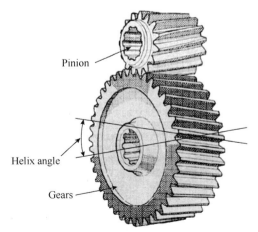

Fig.10.6　Helical gears

10.4 Bevel Gears

Unlike spur and helical gears with teeth cut from a cylindrical blank, bevel gears have teeth cut on an angular or conical surface. Bevel gears are used when input and output shaft centerlines intersect. Teeth are usually cut at an angle so that the shaft axes intersect at 90°, but any other angle may be used. A special class of bevels called miter gears have gears of the same size with their shafts at right angles. Often there is no room to support bevel gears at both ends because the shafts intersect. Thus, one or both gears overhang their supporting shafts. This overhung load (OHL) may deflect the shaft and misalign the gears, which causes poor tooth contact and accelerates wear. Shaft deflection may be overcome with straddle mounting in which a bearing is placed on each side of the gear provided space permits. There are two basic classes of bevels: straight-tooth and spirals.

Straight-tooth bevels. These gears, also known as plain bevels, have teeth cut straight across the face of the gear, Fig.10.7. They are subject to much of the same operating conditions as spur gears in that straight-tooth bevels are efficient but somewhat noisy. They produce thrust loads in a direction that tends to separate the gears.

Spiral-bevels. Curved teeth provide an action somewhat like that of a helical gear, Fig.10.8. This produces smoother, quieter operation than straight-tooth bevels. Thrust loading depends on the direction of rotation and whether the spiral angle at which the teeth are cut is positive or negative.

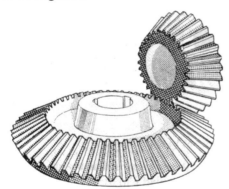
Fig.10.7 Straight-tooth bevel gears

Fig.10.8 Spiral-bevel gears

10.5 Worm Gearing

Worm gear sets, Fig.10.9, consist of a screw-like worm (comparable to a pinion) that meshes with a larger gear, usually called a wheel. The worm acts as a screw, several revolutions of which pull the wheel through a single revolution.[5] In this way, a wide range of speed ratios up to 60 : 1 and higher can be obtained from a single reduction. Most worms are cylindrical in shape with a uniform pitch diameter. However, a double-enveloping worm has a variable pitch diameter that is narrowest in the middle and greatest at the ends. This configuration allows the worm to engage more teeth on the wheel, thereby increasing load capacity. In worm-gear sets, the worm is most often the driving member. However, a reversible worm-gear has the worm and wheel pitches so proportioned that movement of the wheel rotates the worm. In most worm gears, the wheel has teeth similar to those of a helical gear, but the tops of the teeth curve inward to envelop the worm. As a result, the worm slides rather than rolls as it drives the wheel. Because of this high level of rubbing between the worm and wheel teeth, the efficiency of worm gearing is lower than other major gear types. One major advantage of the worm gear is low wear, due mostly to the full-fluid lubricant film that tends to be formed between tooth surfaces by the worm sliding action. Worm-gear shafts are perpendicular, non-intersecting, and may be positioned in a variety of orientations.

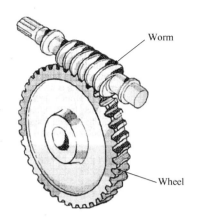

Fig.10.9 Worm gear sets

10.6 Gear Geometry

The fundamentals of gearing are illustrated through the spur gear tooth, both because it is the simplest, and hence most comprehensible, and because it is the form most widely used, particularly for instruments and control systems. The basic geometry and nomenclature of a spur gear mesh is shown in Fig.10.10. The essential features of a gear mesh are:

1. ***Module (m)***: The length, in mm, of the pitch circle diameter per tooth.
2. ***Pressure angle of the contacting involutes*** (α): The angel between the line of force between meshing teeth and the tangent to the pitch circle at the point of mesh.
3. ***Base circle*** (r_b): The circle from which the theoretical curve of gear teeth start.
4. ***Pitch circle (r)***: The circle on a gear on which the thickness of a tooth equals to the space.
5. ***Addendum or outside circle*** (r_a): The circle drawn through the top of the gear tooth, its centre is at the gear centre.
6. ***Root or form circle*** (r_f): The circle drawn through the bottom of the gear tooth, its centre is at the gear centre.
7. ***Addendum*** (h_a): The radial distance from the pitch circle to the addendum circle.
8. ***Dedendum*** (h_f): The radial distance from the pitch circle to the dedendum circle.
9. ***Whole depth (h)***: The radial distance between the addendum and the dedendum circles.
10. ***Working depth***: The radial distance between the addendum and the clearance circles.
11. ***Face width (B)***: The length of teeth in an axial plane.

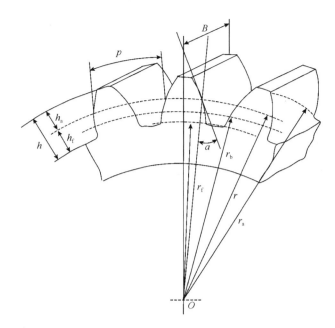

Fig.10.10 Spur gear terminology

12. ***Pitch***: The distance between similar, equally spaced tooth surfaces along a given line or curve.
13. ***Pitch point***: The point of a gear-tooth profile which lies on the pitch circle of that gear. At the moment that the pitch point of one gear contacts its mating gear, the contact occurs at the pitch point of the mating gear, and this common pitch point lies on a line connecting the two gear centers.
14. ***Circular pitch (p)***: The distance measured along the pitch circle, from a point on one tooth to the corresponding point on the adjacent tooth of the gear.
15. ***Diametral pitch***: A measure of tooth size in the English system. In units, it is the number of teeth per inch of pitch diameter. As the tooth size increases, the diametral pitch decreases. Diametral pitches usually range from 1 to 25.
16. *Axial pitch*: Linear pitch in an axial plane and in a pitch surface. In helical gears and worms, axial pitch has the same value at all diameters. In gearing of other types, axial pitch may be confined to the pitch surface and may be a circular measurement.
17. ***Base pitch***: In an involute gear, the pitch on the base circle or along the line of action. Corresponding sides of involute gear teeth are parallel curves, and the base pitch is the constant and fundamental distance between them along a common normal in a plane of rotation.
18. ***Axial base pitch***: The base pitch of helical involute tooth surfaces in an axial plane.
19. ***Tooth surface***: Forms the side of a gear tooth.
20. ***Tooth profile***: One side of a tooth in a cross section between the outside circle and the

root circle.
21. ***Flank***: The working, or contacting, side of the gear tooth. The flank of a spur gear usually has an involute profile in a transverse section.
22. ***Gear center***: The center of the pitch circle.
23. ***Line of centers***: Connects the centers of the pitch circles of two engaging gears; it is also the common perpendicular of the axes in crossed helical gears and worm gears.
24. ***Line of action***: The path of action for involute gears. It is the straight line passing through the pitch point and tangent to the base circle.
25. ***Line of contact***: The line or curve along which two tooth surfaces are tangent to each other.
26. ***Point of contact***: Any point at which two tooth profiles touch each other.
27. ***Center distance***: The distance between the parallel axes of spur gears or of parallel helical gears, or the crossed axes of crossed helical gears or of worms and worm gears. Also, it is the distance between the centers of the pitch circles.
28. ***Lead***: The axial advance of a thread or a helical spiral in 360 deg (one turn about the shaft axis).
29. ***Backlash***: The amount by which a tooth space exceeds the thickness of the engaging tooth on the operating pitch circles.

Gears being an important part of a machine have immense usage within various industries. These industries include automotive industry, coal power plants, steel plants, paper mills, mines and many more. In these industries they behold a wide area of application. They are used as conveyors, elevators, separators, cranes and lubrication systems. The advantages of gear drive over other transmission means are:
1. Gear transmission gives positive drives and constancy of speed ratio without any slippage.
2. Gear drive is very compact due to short centre distances in such drives.
3. Gear transmission has high efficiency, service, and simple operation.
4. Gear drive is capable of driving loads subjected to shock at speeds up to 20 m/s.
5. Maintenance of gear drives is inexpensive and if properly lubricated and operated, gear drives have the longest service life compared to other drives.
6. Gear drive can be used where precise timing is desired.
7. Gear transmission can drive much heavier loads than other drives.
8. Gear drive can be used for a wide range of transmitted power.

Notes

[1] Gears are compact, positive-engagement, power transmission elements in gearing transmission, which determine the speed, torque, and direction of rotation of driven machine elements.

句意：在啮合传动中，齿轮是一种结构紧凑的、强制性啮合的动力传输元件，它可以决定速度、转矩和被驱动的机器零件的旋转方向。

[2] Practice has shown that this design accommodates mostly rolling, rather than sliding, contact of the tooth surfaces.

句意：经验表明，这种设计可以很好地适应齿面的滚动接触，而不是滑动接触。

[3] These rapid shocks result in some objectionable operating noise and vibration.

句意：快速冲击可以导致不良的工作噪声和振动。

[4] Generally, the rack is operated without a sealed enclosure in these applications, but some type of cover may be provided to keep dirt and other contaminants from accumulating on the working surfaces.

句意：上述应用场合中，齿条一般是在没有密封外壳的情况下工作的，但采用外罩可以用来防止在工作表面累积脏物和其他污染物。

[5] The worm acts as a screw, several revolutions of which pull the wheel through a single revolution.

句意：蜗杆可以看成一个螺栓，每转动若干圈就带动涡轮转动一周。

Glossary

accommodate   *v.* 容纳；使适应；为……提供住宿
addendum   *n.* 齿顶高
axial pitch   轴向节距
backlash   *n.* 侧隙
base circle   基圆
base pitch   基节
bevel gear   锥齿轮
circular pitch   周节
dedendum   *n.* 齿根高
diametral pitch   径节
differential   *adj.* 差别的，区别的   *n.* 差别，差异；微分

double-enveloping   二次包络
face width   齿宽
flank   *n.* 侧面
helical gear   斜齿轮
hypoid gear   伞齿轮
machinery   *n.* （总称）机器，机械；机构
miter gear   等径伞齿轮
involute   *n.* 渐开线
lead   *n.* 导程；领先，领导；铅，铅制品   *v.* 引导，指导
line of action   力作用线
lubricant   *adj.* 润滑的   *n.* 润滑物，润滑油，润滑剂
mesh   *n.* 网孔，网丝，网眼；圈套，陷阱；[机]啮合
module   *n.* 模数；组件；模块
nomenclature   *n.* 术语，命名系统
objectionable   *adj.* 不适合的，不能采用的，有害的
overhung load   悬臂载荷
perpendicular   *adj.* 垂直的   *n.* 垂直（线）
pinion   *n.* 小齿轮
pitch   *n.* 节距
pitch circle   节圆（分度圆）
positive-engagement   正啮合
pressure angle   压力角
rack   *n.* 齿条
shaft   *n.* 轴；杆状物；狭长通道
spiral angle   螺旋角
spur gear   直齿轮
straddle   *adj.* 跨式的   *v.* 伸展；跨越
transmission   *n.* 播送，发射；传动，传动装置
trochoidal   *adj.* 摆线的，次摆线的；车轮状的
tooth profile   齿廓
tooth surface   齿面
worm   *n.* 蜗杆

Unit 11　Direct Drive Technology

11.1　Introduction

Direct-drive technology contributes to higher machine (or system) throughput with quicker acceleration and higher top speeds compared with gear-or belt-driven designs.[1] Also, the motor has no brushes to replace. In direct-drive motion, the motor is directly connected to its driven load without intervening ball screws, pulleys, gearboxes, timing belts, or other components. Numerous manufacturers displayed direct-drive motion technology either in the form of low-speed rotary motors or high-speed linear motors.

11.2　Direct-drive Linear (DDL) Motion

This distinct motion technology eliminates all rotary-to-linear conversion devices between motor and load—such as ball screws, gear boxes, rack-and-pinions, and belts, to obtain high-dynamic performance in a growing number of applications.

DDL motors. Core of a typical DDL system is the linear motor. Linear motors are a special class of synchronous brushless servo motor. They work like rotary motors, but are opened up and rolled out flat. Through the electromagnetic interaction between a coil assembly (primary part) and a permanent magnet assembly (secondary part), the electrical energy is converted to linear mechanical energy with a high level of efficiency.[2] Other common names for the primary component are motor, moving part, slider or glider, while the secondary part is also called magnetic way or magnet track. Several linear motor design variants exist.

Three DDL motors. (See Fig.11.1 and Fig.11.2.) Different manufacturing groups have specialized in one or another of three basic linear motor configurations—flat bed, U-channel and tubular. Each motor has its intrinsic advantages and limitations, but drawbacks specific to one motor type can often be sidestepped by using either of the two alternatives.

Flat bed motors, while offering unlimited travel and the highest drive force, exert considerable and undesirable magnetic attraction between the load carrying mover and the motor's permanent magnet track. This attraction force requires bearings that support the extra load.

With its ironless core, the U-channel motor has low inertia and thus maximum agility. However, the mover's load carrying magnetic coils travel deep within the U-Channel frame, restricting heat removal.[3]

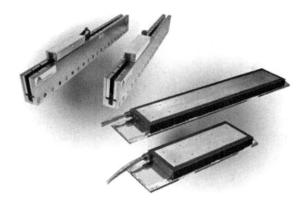

Fig.11.1 Direct-drive ironless linear servo motors (top left), iron-core linear servo motors (lower right)

Fig.11.2 Tubular linear motor

The simplest to install, tubular linear motors are rugged and thermally efficient. Furthermore, they provide drop-in replacements for ballscrew and pneumatic positioners. The tubular motor's permanent magnets are encased in a stainless steel tube (Thrust Rod), which is supported at both ends. Without additional thrust rod support, load travel is limited to 2 to 3 meters, depending on Thrust Rod diameter.

Mature, yet advancing. Bosch Rexroth, Electric Drives and Controls Division, regards DDL motor systems to be a mature technology, yet one that continues to make incremental advancements.[4] The company cites examples, such as improved coolant jacket designs that minimize temperature differential between stator and machine bed, optimized magnet shapes to reduce force ripple and material cost, and magnet tracks protected by stainless steel covers supplied pre-assembled in various lengths, to ease system installation and operation.

Some functions, such as force/torque ripple compensation, nanometer-level interpolation, and ultra-high resolution sinusoidal feedback, are said to result in superior dynamic/static stiffness and motion accuracy.[5] Additionally, automatic commutation functions can eliminate the need for absolute feedback or Hall-effect sensor boxes, although they're supported as well.

Another feature is enhanced vibration suppression, which reduces machine resonance and settling time. It also eliminates vibration due to machine resonance and load disturbances. This function automatically detects and suppresses oscillation frequencies under 1 kHz. A notch filter is also available to control frequencies of 1 kHz and above.

Advantages. Manufacturers have moved beyond specialized semiconductor industry usage to provide advanced performance in a host of applications. In fact, with ten times the speed and ten times the operating life of ballscrews, linear direct drive technology is often the only solution to modern productivity-enhancing automation.

Linear direct drives offer the following advantages over conventional solutions for generating linear motions consisting of a rotary motor and a mechanical power transmission component:

1. Much higher velocities;
2. Much higher acceleration;
3. Much higher positioning accuracy without overshoot;
4. Direct force build-up;
5. Maintenance-free, backlash-free drive;
6. High static and dynamic stiffness under load;
7. Control loop stability due to optimal mass matching.

11.3 Direct-drive Rotary (DDR) Motors Streamline Machine Design

When electric motors are used in low-speed rotary applications, a transmission is often installed to gear down motor speed for higher output torque. This reduces cost because gearing allows small high-speed motors to produce high torque at low-speeds. (See Fig.11.3) However, in high-accuracy applications, such as film-coating lines and integrated circuit test machines, designers avoid gearing because it causes a host of problems like position error, lost motion (backlash), more maintenance, and audible noise.[6]

Most designers would specify so-called direct-drive rotary (DDR) motors (see Fig.11.3), if they could afford them. Until recently, this option is usually reserved for high-end commercial and military applications since direct drive is too expensive for most industrial machines. Fortunately, this has changed.

Direct-drive rotary motors, sometimes called torque motors, develop high torque at relatively low speeds, usually just a few hundred rpm. Two types of DDR motors are available. In frameless motors, customers purchase components such as a rotor, stator, and feedback device. These parts are then assembled with the rest of the machine.

Housed DDR motors integrate the rotor, stator, and feedback device into one assembly. Housed torque motors do not have an intermediate shaft coupling. Instead, the load is attached directly to the DDR rotor. The rotor has a through-hole, typically about 50 mm in diameter, which allows plumbing and wiring to pass through the center. Housed motors have independent bearings, while frameless motors rely on the bearings of the machine.

Fig.11.3   DDR motor

Why use DDR motors? Increasing a machine's precision is the main reason to choose DDR motors. Since the load is rigidly coupled to the motor, error caused by transmission components is eliminated: there is no backlash, belt stretch, or gear-tooth error. The main limitation is the accuracy of the feedback device, but feedback devices for DDR motors are very accurate.

Also, stick-slip is usually eliminated. Stick-slip is a condition in which moving a load over very small distances cannot be done with accuracy.[7] It often comes from transmission components that bring high-friction and high-compliance. Because DDR motors reduce friction and virtually eliminate coupling compliance, they are often not subject to stick-slip.

Another advantage is that the high stiffness between motor and load effectively removes mechanical resonance, a phenomenon in which a compliant load generates instability under high servo gains.[8] This means the servo gains of DDR systems can be set very high, allowing faster servo response and greater resistance to torque disturbances.

Audible noise is also reduced because of fewer moving parts. Maintenance is reduced because the only wearing component in the system is the bearing. If the bearings are permanently lubricated, the assembly can achieve zero maintenance. Machines using DDR motors are often simpler and smaller because the transmission is eliminated. And DDR motors can actually reduce cost in cases where highly accurate transmission components or feedback devices would otherwise be needed.

Not right everywhere. Direct-drive motors are not right for every application. They are usually more costly than conventional rotary motors using transmissions, especially when high gear ratios (>10 : 1) are used to gain mechanical advantage. Feedback devices for DDR motors are usually also more costly. Low friction of DDR-based machine systems, normally an advantage, can be a problem in some designs that rely on friction to bring motion to rest when power is removed.[9]

Finally, for engineers who are familiar with conventional rotary motor design, time is required to learn how to apply DDR technology.

11.4 Motorized Spindle

High speed machining is a promising technology to drastically increase productivity and reduce production costs. The technology of high speed machining is still relatively new. Although theories of high-speed metal cutting were reported in the 1930s, machine tools capable of achieving these cutting speeds did not exist until the 1980s. Only recently, industry has started experimenting with the use of high speed machining in production. The aircraft industry was first, with the automotive industry and mold and die makers now following.

Because of little experience in this new field, there are still many problems to be solved in the application of high speed machining. Current problems include issues of tooling, balancing, thermal and dynamic behaviors, and reliability of machine tools. High speed machining is often associated with high feed rates which require rapid acceleration and deceleration, resulting in drastic changes in cutting conditions. In the aerospace industry the use of long tools to generate deep pockets and thin cross-sections also complicates the problems. Development of high speed spindle technology is strategically critical to the implementation of high speed machining.

Compared with conventional spindles, motorized spindles (see Fig.11.4) are equipped with a built-in motor, so that power transmission devices such as gears and belts are eliminated. This design also reduces vibrations, achieves high rotational balance, and enables precise control of rotational accelerations and decelerations. It soon became obvious that the high speed operation and the very high heat dissipation of built-in motors push other spindle components to their limits. Not only are individual components subject to more severe mechanical and thermal conditions, but spindle design grows more and more complex with added auxiliary components. Eventually, high speed spindles and particularly their thermal and mechanical behavior have become very difficult to calculate efficiently, and experience has become the dominant method for the designer. Nevertheless, experience alone is often insufficient, and modern high speed spindles have serious reliability problems which delay their acceptance in industry.

There is a strong need for better modeling of the thermo-mechanical behavior of an entire motorized high speed spindle system. While programs for the calculation of thermal and mechanical properties of bearings are available and the thermal and mechanical behavior of bearings has been investigated with special test rigs, a comprehensive thermo-mechanical model for motorized high speed spindles does not exist. It is believed that the understanding of thermal and mechanical interactions of different spindle components in practical spindle systems is the key to improving spindle performance and reliability, which cannot be verified satisfactorily in specially designed test rigs for individual components.[10] This is because the isolated effects of individual components can rarely be observed in actual machinery, since several effects occur simultaneously in most cases.

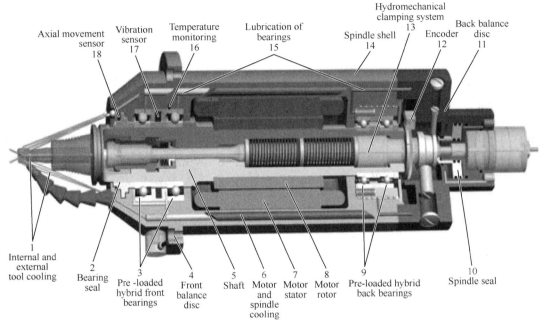

Fig 11.4　Structure of motorized spindle [11]

Notes

[1] Direct-drive technology contributes to higher machine (or system) throughput with quicker acceleration and higher top speeds compared with gear- or belt-driven designs.

句意：直接驱动技术与常规的齿轮、皮带传动相比，由于具有更高的加速度和最高转速，因而可以提高所驱动的机械设备或系统的生产能力。

[2] Through the electromagnetic interaction between a coil assembly (primary part) and a permanent magnet assembly (secondary part), the electrical energy is converted to linear mechanical energy with a high level of efficiency.

句意：通过线圈组（初级）和永磁体（次级）之间的电磁感应，电能可以高效地转化为线性运动机械能。

[3] However, the mover's load carrying magnetic coils travel deep within the U-Channel

frame, restricting heat removal.

句意：然而，由于动子的负载磁力线圈是在 U 形槽框架的深槽内移动，因此限制了热量的散发。

[4] Bosch Rexroth, Electric Drives and Controls Division, regards DDL motor systems to be a mature technology, yet one that continues to make incremental advancements.

句意：德国博世-力士乐公司的电力驱动和控制部认为，DDL 电机系统是成熟的技术，但也是有待不断持续发展进步的技术。

[5] Some functions, such as force/torque ripple compensation, nanometer-level interpolation, and ultra-high resolution sinusoidal feedback, are said to result in superior dynamic/static stiffness and motion accuracy.

句意：有些功能，比如力/力矩脉动补偿、纳米级插补，以及超高分辨率正弦反馈，可以产生优良的动态/静态刚度和运动精度。

[6] However, in high-accuracy applications, such as film-coating lines and integrated circuit test machines, designers avoid gearing because it causes a host of problems like position error, lost motion (backlash), more maintenance, and audible noise.

句意：然而，在诸如镀膜线和集成电路测试仪等高精度应用场合，设计者尽量不用传动装置，因为它会带来很多问题，比如定位误差、空转（齿侧间隙）、频繁维护及音频噪声。

[7] Also, stick-slip is usually eliminated. Stick-slip is a condition in which moving a load over very small distances cannot be done with accuracy.

句意：同样地，爬行现象也消除了。爬行是这样一种现象：不能够做到在很小的距离内准确移动负载。

[8] Another advantage is that the high stiffness between motor and load effectively removes mechanical resonance, the phenomenon in which a compliant load generates instability under high servo gains.

句意：另一个好处是，电机和负载之间的高刚度有效去除了机械共振现象，这种现象中，相容负载在高伺服增益时会产生不稳定性。

[9] Low friction of DDR-based machine systems, normally an advantage, can be a problem in some designs that rely on friction to bring motion to rest when power is removed.

句意：通常认为 DDR 机械系统的低摩擦力是一个优点，但在某些设计中却是一个问题，在这些设计中，当停电时要依靠摩擦力使运动停止。

[10] It is believed that the understanding of thermal and mechanical interactions of different spindle components in practical spindle systems is the key to improving spindle

performance and reliability, which cannot be verified satisfactorily in specially designed test rigs for individual components.

句意：人们相信，在实际的主轴系统中，对不同主轴部件的热和机械相互作用的了解，是提高主轴的性能和可靠性的关键，但在对单个部件专门设计的测试平台上，这一观点无法得到令人满意的证实。

[11] 1. Internal and external tool cooling 刀具冷却系统；2. Bearing seal 轴承密封装置；3. Pre-loaded hybrid front bearings 预载荷动静压前轴承；4. Front balance disc 前平衡盘；5. Shaft 转轴；6. Motor and spindle cooling 主轴和电机冷却装置；7. Motor stator 电机定子；8. Motor rotor 电机转子；9. Pre-loaded hybrid back bearings 预载荷动静压后轴承；10. Spindle seal 主轴密封装置；11. Back balance disc 后平衡盘；12. Encoder 编码器；13. Hydromechanical clamping system 液压夹紧装置；14. Spindle shell 主轴外壳；15. Lubrication of bearings 轴承润滑装置；16.Temperature monitoring 温度监测装置；17. Vibration sensor 振动传感器；18. Axial movement sensor 轴向位移传感器。

Glossary

agility　*n*. 敏捷，灵活性
audible　*adj*. 听觉的，听得见的
auxiliary　*adj*. 辅助的，补助的，补充的，副的，附属
backlash　*n*. 间隙，齿隙，反向间隙
ballscrew　*n*. 滚珠丝杠
commutation　*n*. 换向，转接，切换
compliant load　相容负载
deceleration　*n*. 减速度，制动，熄灭，负加速
direct-drive linear (DDL) motor　直接驱动直线电机
direct-drive rotary (DDR) motor　直接驱动旋转电机
dominant　*adj*. 支配的，统治的
drop-in replacement　快插式更换（零件）
dynamic behavior　动态性能
frameless motor　无外壳电机
friction　*n*. 摩擦，摩擦力
gearbox　*n*. 齿轮箱，变速箱，进刀箱，进给箱，减速箱
Hall-effect　霍尔效应
housed DDR motor　封装式直接驱动旋转电机
incremental　*adj*. 增量的，逐渐增长的，递增的

integrate　*v.* 使结合，使并入，使一体化

interpolation　*n.* 插值，内插，内插法

intrinsic　*adj.* 固有的，本身的，内在的

ironless　*n.* 无铁的，无铁心的

load carrying forcer　负载平台

lubricate　*v.* 使润滑，加润滑油

mover　*n.* 动子

motorized spindle　电主轴

nanometer-level　纳米级

oscillation　*n.* 动摇，摆动，振荡；振幅，消长度，上下波动

overshoot　*vt.* 超调，过冲

permanent magnet track　永磁轨道

pneumatic　*adj.* 空气的，气动的

positioner　*n.* 定位器

resonance　*n.* 共振

ripple　*n.* 波动，变化

rotary　*adj.* 旋转的，转动的，轮转的，循环的，轮流的

rugged　*adj.* 粗壮，结实的

servo　*n.* 伺服机构，伺服电机

sidestep　*v.* 回避，逃避

simultaneously　*adv.* 同时地

sinusoidal　*adj.* 正弦曲线的

stick-slip　（机床）爬行现象

synchronous　*adj.* 同时的，同期的，同步的

thermal　*adj.* 热的

thrust rod　推（力）杆

timing belt　同步带，齿形皮带

torque motor　力矩电机

tubular　*adj.* 管形的，筒形的

ultra-high　超高的

vibration　*n.* 振动，摆动，振荡

Unit 12   Numerical Control

12.1   Introduction

One of the most fundamental concepts in the area of advanced manufacturing technologies is numerical control (NC). Prior to the advent of NC, all machine tools were manually operated and controlled. Among the many limitations associated with manual control machine tools, perhaps none is more prominent than the limitation of operator skills. With manual control, the quality of the product is directly related and limited to the skills of the operator. Numerical control represents the first major step away from human control of machine tools.

12.2   NC and CNC

Numerical control means the control of machine tools and other manufacturing systems through the use of prerecorded, written symbolic instructions. Rather than operating a machine tool, an NC technician writes a program that issues operational instructions to the machine tool. An NC machine tool can automatically produce a wide variety of parts, each involving an assortment of widely varied and complex machining processes. Numerical control has allowed manufacturers to undertake the production of products that would not have been feasible from an economic perspective using manually controlled machine tools and processes.

The original NC systems were vastly different from those used today. The machines had hardwired logic circuits. The instructional programs were written on punched paper, which was later replaced by magnetic plastic tape. A tape reader was used to interpret the instructions written on the tape for the machine. The development of a concept known as direct numerical control (DNC) solved the paper and plastic tape problems associated with numerical control by simply eliminating tape as the medium for carrying the programmed instructions.[1] In direct numerical control, machine tools are tied, via a data transmission link, to a host computer. Programs for operating the machine tools are stored in the host computer and fed to the machine tool as needed via the data transmission linkage. Direct numerical control represented a major step forward over punched tape and plastic tape. However, it is subject to the same limitations as all technologies that depend on a host computer. When the host computer goes down, the machine tools also experience downtime. This problem led to the development of computer numerical control.

The development of the microprocessor allowed for the development of programmable logic controllers (PLCs) and microcomputers.[2] These two technologies allowed for the development of computer numerical control (CNC). With CNC, each machine tool has a PLC or a microcomputer that serves the same purpose. This allows programs to be input and stored at each individual machine tool. It also allows programs to be developed off-line and downloaded to the individual machine tool. CNC solved the problems associated with downtime of the host computer, but it introduced another problem known as data management. The same program might be loaded to ten different microcomputers with no communication among them. This problem is in the process of being solved by local area networks that connect microcomputers for better data management.

12.3 Construction of CNC Machines

CNC machine tools are complex assemblies. In general, any CNC machine tool consists of the following units:

1. Computers;
2. Control systems;
3. Drive motors;
4. Tool changers.

According to the construction of CNC machine tools, CNC machines work in the following (simplified) manner:

1. The CNC machine control computer reads a prepared program and translates it into machine language, which is a programming language of binary notation used on computers, not on CNC machines.
2. When the operator starts the execution cycle, the computer translates binary codes into electronic pulses which are automatically sent to the machine's power units. The control units compare the number of pulses sent and received.
3. When the motors receive each pulse, they automatically transform the pulses into rotations that drive the spindle and lead screw, causing slide or table movement. The part on the milling machine table or the tool in the lathe turret is driven to the position specified by the program.[3]

Computers. CNC machines use an on-board computer that allows the operator to read, analyze, and edit programmed information, while NC machines require operators to make a new tape to alter a program. In essence, the computer is what distinguishes CNC from NC.

As with all computers, the CNC machine computer works on a binary principle using only two characters, 1 and 0, for information processing. The computer reacts on precise time impulses from the circuit. There are two states, a state with voltage, 1 and a state without voltage, 0. Series of ones and zeroes are the only states that the computer distinguishes, called *machine language*, it is the only language the computer understands. When creating the program, the programmer does not care about the machine language; he or she simply uses a list of codes and keys in the meaningful information. Special built-in software compiles the program into machine language and the machine moves the tool by its servomotors.[4]

Modern CNC machine use 64-bit processors in their computers that allow fast and accurate processing of information. This results in savings of machining time.

Control systems. There are two types of control systems on NC/CNC machines: open-loop and closed-loop. The overall precision of the machine is determined by the type of control loop used.

The open-loop control system does not provide positioning feedback to the control unit. The movement pulses are sent out by the control and received by a special type of servomotor called a stepping motor. The number of pulses that the control sends to the stepping motor controls the amount of the rotation of the motor. The stepping motor then proceeds with the next movement command. Since this control system only counts pulses and cannot identify the discrepancies in positioning, the control has no way of knowing that the tool did not reach the proper location. The machine will continue this inaccuracy until somebody finds the error.

The open-loop control can be used in applications in which there is no change in load conditions, such as NC drilling machine. The advantage of the open-loop control system is that it is less expensive, since it does not require the additional hardware and electronics needed for positioning feedback. The disadvantage is the difficulty of detecting a positioning error.

In the closed-loop control system, the electronic movement pulses are sent from the control to the servomotor, enabling the motor to rotate with each pulse. The movements are detected and counted by a feedback device called *transducer*. With each step of movement, a transducer sends a signal back to the control, which compares the current position of the driven axis with the programmed position. When the numbers of pulses sent and received match, the control starts sending out pulses for the next movement.

Closed-loop systems are very accurate. Most have an automatic compensation for error, since the feedback device indicates the error and the control makes the necessary adjustments to bring the slide back to the position. They use AC, DC, or hydraulic servomotors.

Drive motors. The drive motors control the machine slide movement on NC/CNC equipment.

They come in four basic types:

1. Stepping motors;
2. DC servomotors;

3. AC servomotors;

4. Fluid servomotors.

Stepping Motors convert a digital pulse, generated by the microcomputer unit (MCU), into a small step rotation. Stepping motors have a certain number of steps that they can travel. The number of pulses that the MCU sends to the stepping motor controls the amount of the rotation of the motor. Stepping motors are mostly used in applications where low torque is required.

Stepping motors are used in open-loop control systems, while AC, DC, or hydraulic servomotors are used in closed-loop control systems.

Direct current (DC) servomotors are variable-speed motors that rotate in response to the applied voltage. They are used to drive a lead screw and gear mechanism. DC servos provide higher-torque output than stepping motors.

Alternative current (AC) servomotors are controlled by varying the voltage frequency to control speed. They can develop more power than a DC servo. They are also used to drive a lead screw and gear mechanism.

Fluid, or hydraulic, servomotors are also variable-speed motors. They are able to produce more power, or more speeds in the case of pneumatic motors, than electric servomotors.

Tool changers. Most of the time, several different cutting tools are used to produce a part. The tool must be replaced quickly for the next machining operation. For this reason, the majority of NC/CNC machine tools are equipped with automatic tool changers, such as magazines on machining centers and turrets on turning centers (see Fig.12.1).They allow tool changing without the intervention of the operator. Typically, an automatic tool changer grips the tool in the spindle, pulls it out, and replaces it with another tool. On most machines with automatic tool changers, the turret or magazine can rotate in either direction, forward or reverse.

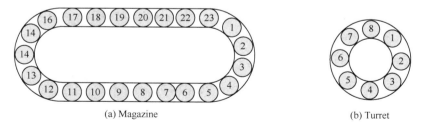

(a) Magazine   (b) Turret

Fig.12.1 The automatic tool changers used most often

12.4 DNC (Distributed Numerical Control) System

With the development of computer and communication technology, the DNC connotation has transformed from direct numerical control to distributed numerical control. Compared with

FMS (flexible manufacturing system), DNC, which has higher efficiency and needs less investment, doesn't emphasize materials flow automation.[5] It is being used widely day by day and is becoming a research focus as well.

At present, a few NC machine tools have Ethernet, MAP network of field bus interfaces, a majority of NC machine tools have only RS232, RS485 or RS422 communication interfaces. There are two kinds of DNC which are often used to communicate with NC machine tools.

Star structure DNC. Because of the high price of computers and other reasons, star structure DNC which was made up of several NC machine tools linked with a single DNC computer, the main controller of DNC, predominated from the end of 1980s to the beginning of 1990s.

Although such DNC overcomes the investment problem and also has a simple structure, it has several disadvantages: (1) the NC machine tools networked must be relatively centralized; (2) the whole system will be paralyzed if the DNC computer breaks down; (3) there are too many lines between RS232 interfaces, and the reliability is very low; and (4) the overload of DNC computer often results in communication competition.

Local area network (LAN) plus point-to-point DNC. The star structure DNC cannot satisfy the requirements of nowadays production, because more and more NC machine tools are installed dispersedly. LAN plus point-to-point DNC is another kind of DNC used widely, in which one DNC computer connects to one NC machine tool, and all DNC computers are linked to the CAD/CAM system by LAN. This kind of DNC is simple to link, and can adjust the quantity of NC machine tools easily. Its communication rate is high, suitable for a large number of NC machine tools installed dispersedly. Its network topological structure is shown in Fig.12.2.

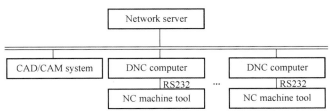

Fig.12.2　LAN plus point-to-point DNC

The essence of above-mentioned DNC is that a DNC computer is connected to a NC machine tool through an RS232 interface; the distance of communication is short and the DNC computer must be located at the production spot. Such DNC has the following defects: (1) it is inconvenient to operate, for the NC machine tool and the DNC computer need to be manipulated separately; (2) the communication software needs to be developed according to different numerical control systems; and (3) the DNC computer at the production site is inevitably subject to dust and environmental temperature, which results in a high failure rate of the DNC computer, and hence incurs frequent interruption of production and increased maintenance cost as well. For this reason, it is necessary to develop a novel DNC integrated system that is a real-time, high-speed, and reliable remote communication system. In addition, the system must be convenient to operate, and has good expansibility.

CAN Bus—based structure for DNC system. The hierarchical control architecture may underpin many solutions of integrated factory control, where decisions are made at the top-level controllers, and the lower levels simply execute the command. In this situation, two hierarchical layers of network are generally conducted, which can be easily involved in the automatic system of workshop-floor and also can be expanded as a large-scale manufacturing system, including planning, design, production, management, sale and service. The structure is shown in Fig.12.3.

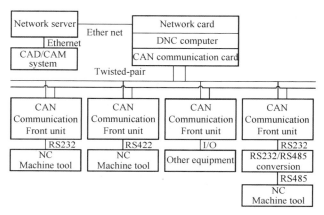

Fig.12.3 Hierarchical structure of the automation manufacturing workshop based CAN Bus

The top layer is TCP/IP based Ethernet, such as Windows NT system of Novell, copes with CAD/CAPP/CAM/PDM. So the main control functions of workshop-floor of factory of enterprise are involved in this layer, which can actually be extended to contain a lot of functional components of manufacturing system, such as production management, design, quality control etc.

The bottom is CAN Bus (Controller Area Network Bus) based field bus system. Almost all kinds of digital equipment, such as NC machine tools, coordinate measuring machines (CMMs), facility of cutting tool pre-adjusting and so on, can be effectively connected with the CAN Bus via two-twisted wire and the special-developed CAN front units, which are also able to be installed in the cabinet of machine tools individually.[6] Because of using verified protocol, the DNC system is robust enough and can run without any malfunction for a long time, at least one year in practice, which means having the ability of immunity to unexpected signal disturbance. And the computer connected to machining equipment as DNC server can be located in a cleaning environment equipped with air-conditioner, which can avoid unforeseen interruption of the machining process because of the computer's breakdown. Additionally, it is quite cheap but reliable that all the device-level facilities are linked by two-twisted wire.

As device-level network, also called field network, CAN Bus mainly deals with delivering information of production, such as NC code, parameters of tools, etc, to the digital facilities involved in this network, and also collecting the data from machining process. On the other hand, the handling method is actually welcomed by the operators because they just need to pass their requirements to others and get the necessary data from DNC server or the top-level network directly on the panel of machine tools. Obviously, the redoing operations both on the keyboard of computer and panel of machine tool are therefore easily dodged.

Additionally, a CAN interface card for PC, inserted in one computer which is the server of DNC system, acts as a bridge to link the two-layer networks together.

Notes

[1] The development of a concept known as direct numerical control (DNC) solved the paper and plastic tape problems associated with numerical control by simply eliminating tape as the medium for carrying the programmed instructions.

句意：在形成直接数字控制（DNC）这个概念之后，可以不再采用纸带或塑料带作为编程指令的载体，这样就解决了与之相关的问题。

[2] The development of the microprocessor allowed for the development of programmable logic controllers (PLCs) and microcomputers.

句意：微处理器的发展为可编程逻辑控制器和微型计算机的发展做好了准备。

[3] The part on the milling machine table or the tool in the lathe turret is driven to the position specified by the program.

句意：由程序驱动铣床台面上的零件或车床回转头上的刀具至指定位置。

[4] Special built-in software compiles the program into machine language and the machine moves the tool by its servomotors.

句意：专用的内置编程软件将程序汇编成机器码，由伺服电机驱动机床使刀具运动。

[5] Compared with FMS (flexible manufacturing system), DNC, which has higher efficiency and needs less investment, doesn't emphasize materials flow automation.

句意：与FMS（柔性制造系统）相比，DNC不强调物流自动化，有更高的效率，需要较少的投资。

[6] Almost all kinds of digital equipment, such as NC machine tools, coordinate

measuring machines (CMMs), facility of cutting tool pre-adjusting and so on. can be effectively connected with the CAN Bus via two-twisted wire and the special-developed CAN front units which are also able to be installed in the cabinet of machine tools individually.

句意：几乎所有的数字设备，比如 NC 机床、三坐标测量机（CMMs）、切削刀具预调整设备等都可以通过双绞线和 CAN 总线专用前端控制单元有效地连入 CAN 总线，这些前端控制单元也可独立安装在机床的控制柜中。

Glossary

advent   *n.* 到来，出现，降临
built-in   内置
centralize   *v.* 集中
connotation   *n.* 含蓄；内涵
discrepancy   *n.* 偏差
dispersedly   *adv.* 分散的，散开的
dodge   *v.* 避开，躲避
field bus   现场总线
hierarchical   *adj.* 分级的，分层的
immunity   *n.* 免疫性，抗干扰能力
interface   *n.* 接口，界面
malfunction   *n.* 故障，失灵
on-board computer   机载计算机
paralyze   *v.* 使瘫痪，使麻痹
predominate   *v.* 掌握，控制
prerecord   *v.* 预先录制
proceed with   继续进行
pneumatic   *adj.* 气动的，由压缩空气推动的
turret   *n.* 回转头
underpin   *v.* 支持；巩固，加强……的基础
unforeseen   *adj.* 无法预料的，意料之外的
via   *prep.* 经过，通过，借助于
workshop-floor   车间，工厂

Unit 13 CNC Machining Centers

13.1 Introduction

In 1968, an NC machine was marketed, which could automatically change tools, so that many different processes could be done on one machine. Such a machine became known as a "machining center"—a multifunctional CNC machine that can perform a variety of processes and change tools automatically while under programmable control. The computer-controlled machining centers have the required flexibility and versatility that other individual machine tools do not have, so they often become the first choice in machine-tool selection.

In describing the individual machining processes and machine tools in Unit 5, it was noted that each machine, regardless of how highly it is automated, is designed to perform basically the same type of operations, such as turning, boring, drilling, milling, broaching, planing, or shaping. Note, for example, that the parts shown in Fig.13.1 have a variety of complex geometric features and that all of the surfaces on these parts require different types of machining operations (such as milling, facing, boring, drilling, reaming, or threading) to meet certain specific requirements concerning shapes, features, dimensional tolerances, and surface finish.

Fig.13.1 The motorcycle wheel with complex geometric features

If several forms of machining are required or if it is needed to finish machining these parts to their final shapes more economically, then it is obvious that none of the machine tools described in Unit 5 individually could produce these parts completely. We also note that, traditionally, machining operations are performed by moving the workpiece from one machine tool to another until all of the required machining operations are completed.

13.2 The Concept of Machining Centers

The traditional method of machining parts using different types of machine tools has been and continues to be a viable and efficient manufacturing method. It can be highly automated in order to increase productivity, and it is indeed the principle behind transfer lines (also called dedicated manufacturing lines, DML). Commonly used in high-volume or mass production, transfer lines consist of several specific machine tools arranged in a logical sequence.

The workpiece (such as an automotive engine block) is moved from station to station with a specific machining operation performed at each station, after which it is transferred to the next machine for another specific machining operation, and so on. There are situations, however, where transfer lines are not feasible or economical, particularly when the types of products to be processed are changed rapidly due to factors such as product demand or changes in product shape or style. It is a very expensive and time-consuming process to rearrange these machine tools to respond to the needs for the next production cycle. An important concept developed in the late 1950s is that of machining centers.

A machining center is an advanced, computer-controlled machine tool that is capable of performing a variety of machining operations on different surfaces and different orientations of a workpiece without having to remove it from its workholding device or fixture. The workpiece generally is stationary, and the cutting tools rotate as they do in milling, drilling, honing, tapping, and similar operations. Whereas in transfer lines or in typical shops and factories the workpiece is brought to the machine, note that in machining centers, it is the machining operation that is brought to the workpiece.

In referring to the word workpiece, we also should point out that the workpiece in a machining center also includes all types of tooling. Tooling can include forming and cutting tools, cutter and tool holders, tool shanks for holding tool inserts, molds for casting, male and female dies for forming, punches for metal-working and powder metallurgy, rams for extrusion, workholding devices, gages and fixtures, and other accessories—all of which also have to be manufactured.[1] Since the geometries are often quite complex and a variety of machining operations are performed, these tools are produced commonly in machining centers.

13.3 Types of Machining Centers

There are various designs for machining centers. The two basic types are the vertical type and the horizontal type. Their names are derived from their respective spindle designs, which are either in a vertical position or horizontal position. There are other designs and variations of both vertical and horizontal machines that are designed for other application needs.

Vertical-spindle machining centers. Also called vertical machining centers (VMC). The

VMC spindle holds the cutting tool in a vertical position. VMCs are generally used to perform operations on flat parts that require cutting on the top surface of the part and on parts with deep cavities, such as in mold and die making. A vertical-spindle machining center (which is similar to a vertical-spindle milling machine) is shown in Fig.13.2. The tool magazine is on the left of the figure, and all operations and movements are controlled and modified through the computer-control panel shown on the right. The VMC is programmed to position and cut in the X-, Y- and Z-axes (three-axes). Other options can be added that will increase the flexibility and productivity of the VMC. Some of the options that are typically added to VMCs include an indexer or a shuttle table. The indexer, which is mounted on the machine table, can rotate the part along the X-axis for 360° machining (fourth axis). The shuttle table option is used for shuttling the parts, which allows the operator to load and unload parts without interrupting production. Because the thrust forces in vertical machining are directed downward, such machines have high stiffness and produce parts with good dimensional accuracy. These machines generally are less expensive than horizontal-spindle machines.

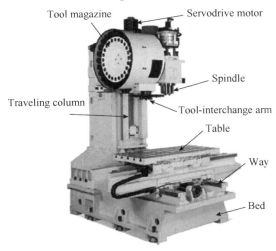

Fig.13.2　Vertical-spindle machining center

Horizontal-spindle machining centers. Also called horizontal machining centers (HMC). The HMC spindle holds the cutting tool in a horizontal position, which is suitable for large as well as tall workpieces that require machining on a number of their surfaces (see Fig.13.3). The HMC machine table can be programmed to rotate 360° in a circular motion (B-axis), so it is programmed to position and cut in the X-, Y-, Z- and B-axes (four-axes). The HMCs can machine parts on more than one side in one clamping, and find wide use in flexible manufacturing systems. The pallet can be swiveled on different axes to various angular positions. Another category of horizontal-spindle machines is turning centers, which are computer-controlled lathes with several features. A three-turret turning center is shown in

Fig.13.4. It is constructed with two horizontal spindles and three turrets equipped with a variety of cutting tools used to perform several operations on a rotating workpiece.

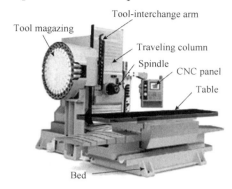

Fig.13.3　Horizontal-spindle machining center

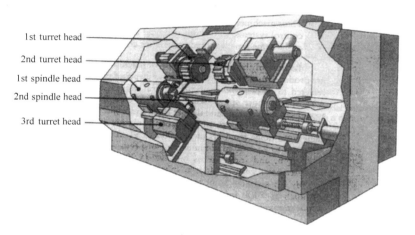

Fig.13.4　Schematic illustration of a CNC turning center

Universal machining centers are equipped with both vertical and horizontal spindles. They have a variety of features and are capable of machining all of the surfaces of a workpiece (that is, vertical and horizontal and at a wide range of angles).

13.4　Components of a Machining Center

CNC machining centers have the same basic components that conventional milling machines have, such as the main motor, the spindle, the table, the frame, and the way system. The CNC machining center, in addition, has a computerized control and servomotors to operate the machine. Thus, the vertical and horizontal machining centers are numerically controlled milling machines that change tools, position the spindle, and cut material automatically. When milling-type machines are coupled with CNC, they become the "high-tech production CNC centers" of the machining shop. Drilling, tapping, reaming, milling, and boring are perfect machining applications for CNC machining centers.

The main components of CNC machining centers are the frame, column, spindle, table, tool magazine, operator control panel, servomotors, ball screws, hydraulic and lubrication systems, and the MCU (Machine Control Unit).

Frame. The frame supports and aligns the axis and cutting tool components of the machine. The frame will absorb the shock and vibration associated with metal cutting conditions. Frames are designed in one of two ways: either cast iron or fabricated.[2] Most machining centers have a fabricated frame design.

Headstock. The headstock contains the spindle and transmission gearing, which rotate the cutting tool. The headstock spindle is driven by a variable-speed motor. Headstocks are equipped with a variety of motor capacities ranging from 5 to 30 horsepower, and spindle speeds from 32 to 10000 rpm.

Tool magazine. The basic function of the tool magazine is to hold and quickly index the CNC cutting tools. Tool magazines are designed in various styles and sizes. CNC machining centers are usually equipped with a tool magazine that can hold a range of twenty to over one hundred tools.[3] The tool magazine can automatically change the cutting tools with the tool change arm. In order to increase productivity, most tool magazines are capable of bidirectional movement to select the tools using the quickest path.

Automatic tool change arm. The automatic tool change (ATC) arm is designed to grasp and then remove the CNC tool from the tool magazine and insert it into the taper bore of the spindle. The ATC arm must accurately align each tool with the bore and the drive keys. Tools are identified by bar codes, coded tags, or memory chips attached directly to their toolholders.[4] The ATC arm can complete a tool change in either manual or automatic mode in approximately 5 to 10 seconds, but may be up to 30 seconds for tools weighing up to 110 kg. Because tool changing is a non-cutting operation, the continuing trend is to reduce the time even more.

The tool-exchange arm shown in Fig.13.5 is a common design; it swings around to pick up a particular tool and places it in the spindle. Note that each tool has its own toolholder, thus making the transfer of cutting tools to the machine spindle very efficient.

Table/Pallet. The table is designed to hold the workholding device, which holds the workpiece while it is being machined. The workholding device, such as a milling machine vise or a fixture, is clamped to the table using bolts and nuts. The table can be programmed or manually operated to move in the *X*- and *Y*-axes. Additionally, the table for HMCs can be rotated about the *B*-axis. The workpiece in a machining center is placed on a pallet or module that can be moved and swiveled (oriented) in various directions.[5] After a particular machining operation has been completed, another operation begins, which may require reindexing of the workpiece on its pallet. After all of the machining operations have been completed, the pallet automatically moves away with the finished part, and another pallet (carrying another workpiece to be machined) is brought into position by an automatic pallet changer (Fig.13.6). All movements are computer controlled, and pallet-changing cycle times are on the order of only 10 to 30 seconds. Pallet stations are available with several pallets serving the machining center. The machines can also be equipped with various automatic features, such as part loading and unloading devices.

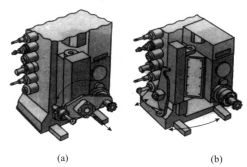

Fig.13.5　An automatic change arm on a horizontal-spindle center

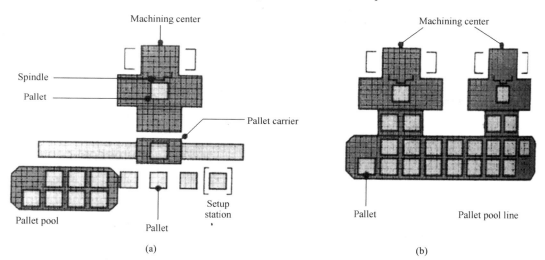

Fig.13.6　Schematic illustration of the top view of a HMC with pallets

Servodrive motors and ball screw. CNC machine use electric servomotors that turn ball screws, which in turn drive the different axes of the machine tool. Each axis has a separate servomotor and ball screw to control the X-, Y-, and Z-axis independent of each other.

The rotary motion generated by the drive motors is converted to linear motion by recirculating ball screws. The ball lead screw uses rolling motion rather than the sliding motion of a normal lead screw. Sliding motion is used on conventional trapezoidal lead screws.[6] Unlike the ball screw, the motion principle of a trapezoidal lead screws is based on friction and backlash. Here are some advantages of the ball screw over the trapezoidal lead screws:

1. Less wear;　　　　　　　　　　　　　2. High-speed capability;

3. Precise position and repeatability; 4. Long life.

13.5　Characteristics and Capabilities of Machining Centers

The major characteristics of machining centers are summarized here:

1. Machining centers are capable of handling a wide variety of part sizes and shapes efficiently, economically, repetitively, and with high dimensional accuracy-with tolerances in the order of ±0.0025 mm.
2. These machines are versatile and capable of quick change-over from one type of product to another.
3. The time required for loading and unloading workpieces, changing tools, gaging of the part, and troubleshooting is reduced. Therefore productivity is improved, thus reducing labor requirements (particularly skilled labor) and minimizing production costs.
4. These machines are equipped with tool-condition monitoring devices for the detection of tool breakage and wear as well as probes for tool-wear compensation and tool positioning.
5. In-process and post-process gaging and inspection of machined workpieces are now features of machining centers.
6. These machines are relatively compact and highly automated and have advanced control systems, so one operator can attend to two or more machining centers at the same time, thus reducing labor costs.

Machining centers are available in a wide variety of sizes and features, and their costs range from about $50000 to 1 million and higher. Typical spindle speed is 8000 rpm (round per minute), and sometime can be as high as 75000 rpm for special applications using small-diameter cutters. Modern spindle can accelerate to a speed of 20000 rpm in only 1.5 seconds. Some pallets are capable of supporting workpieces weighing as much as 7000 kg, although even higher capacities are available for special applications.

Notes

[1] Tooling can include forming and cutting tools, cutter and tool holders, tool shanks for holding tool inserts, molds for casting, male and female dies for forming, punches for metal-working and powder metallurgy, rams for extrusion, workholding devices, gages and fixtures, and other accessories —all of which also have to be manufactured.

句意：工（艺）装（备）包括成形和切削工具，刀具和工具的固定装置（如刀架），用于紧固刀具硬质刀头的刀柄，铸模，成形锻压的凸模和凹模，金属加工和粉末冶金用的冲头，挤压柱塞，工件夹紧装置，还有量具和夹具及其他辅具等——它们都需要制造出来。

[2] Frames are designed in one of two ways: either cast iron or fabricated.

句意：机架可以基于两种结构方式进行设计：铸铁或加工装配式。

[3] Tool magazines are designed in various styles and sizes. CNC machining centers are usually equipped with a tool magazine that can hold a range of twenty to over one hundred tools.

句意：刀库可以设计成不同的形式和大小，CNC 加工中心通常配备有可以容纳 20 至 100 多把刀具的刀库。

[4] Tools are identified by bar codes, coded tags, or memory chips attached directly to their toolholders.

句意：通过条形码、编码标签和直接与刀柄连接的记忆芯片可以实现刀具的识别。

[5] The workpiece in a machining center is placed on a pallet or module that can be moved and swiveled (oriented) in various directions.

句意：加工中心的工件一般放置在可以沿不同方向移动或旋转的托盘上。

[6] The ball lead screw uses rolling motion rather than the sliding motion of a normal lead screws. Sliding motion is used on conventional trapezoidal lead screws.

句意：滚珠丝杠是利用滚动而不像标准的丝杠是利用滑动进行工作的。滑动一般用于传统的梯形丝杠。

Glossary

automatic tool change (ATC)　自动换刀装置
backlash　*n*. 齿侧间隙，螺纹间隙
ball screw　滚珠丝杠
boring　*n*. 镗孔
broaching　*n*. 拉削；扩孔
cast iron　铸铁
characteristics　*n*. 特征，特性
column　*n*. 立柱
control panel　控制面板
dedicated manufacturing lines　专业生产线
desirability　*n*. 客观需求；愿望
drawbar　*n*. 拉杆；导杆
emphasize　*v*. 强调；着重
evolve　*v*. 使发展成；使形成；演化；进展
extrusion　*n*. 挤出；推出；压出

feasible　*adj.* 可能的，切实可行的
female die　阴模
flexibility　*n.* 柔性，弹性；灵活性
frame　*n.* 机架
gaging　*n.* 测量
gasket　*n.* 衬垫；垫圈；密封垫
headstock　*n.* 主轴箱
high-volume or mass production　大（批）量生产
horizontal machining center　卧式加工中心
horsepower　*n.* 马力
hydraulic and lubrication system　液压和润滑系统
indexer　*n.* 分度器
individual　*n.* 个人，个体，独立单位　*adj.* 个人的；个别的；独特的
male die　阳模
machine control unit (MCU)　机床控制单元
net-shape　终形
observation　*n.* 执行，遵守；（观察所得的）知识；经验；观察值
orientation　*n.* 定位；方向
pallet　*n.* 托盘
powder metallurgy　粉末冶金
precision forging　精锻
radii　*n.* (*pl*) 半径
ram　*n.* 冲压；滑枕；柱塞
rearrange　*v.* 再排列；重新整理
refixture　*v.* 二次装卡
regardless of　不顾，不惜，不注意
retention knob　止动旋钮
servomotor　*n.* 伺服电机
shock　*n.* 冲击；震动　*v.* （使）震动
shuttle table　移动工作台
station　*n.* 站；观察站；装配站
stationary　*adj.* 静止的，不动的
stiffness　*n.* 刚度
surface finish　表面光洁度
swivel　*v.* （使）旋转
taper　*n.* 坡度，锥度；锥体　*v.* （使）一头逐渐变细，（使）逐渐减少

tool magazine　刀库

threading　*n.* 车螺纹

turning center　车削中心

universal machining center　通用/万能加工中心

versatility　*n.*（才能、用途等）多面性；通用性；适能力

vertical machining center　立式加工中心

viable　*adj.* 可行的，可能成功的

vibration　*n.* 振动；摆动；振荡

Unit 14　Automatic Control

14.1　Introduction

If we examine the word control, we find several meanings given in the dictionary, e.g., command, direct, govern, and regulate. Thus, a control system may be regarded as a group of physical components arranged to direct the flow of energy to a machine or process in such a manner as to achieve the desired performance.

The word automatic means self-moving or self-acting; thus an automatic control system is a self-acting control system.

An important distinction applied to control systems, whether automatic or otherwise, is that between open-loop and closed-loop operation. Automatic control, including this distinction, can perhaps be best introduced by means of a simple example.

14.2　Open-loop Control and Closed-loop Control

In an example of water cistern, if we want to maintain the actual water level c in the tank as close as possible to a desired level r which is called the system input, and the actual water level is the controlled variable or system output. Water flows from the tank via a valve V_o and enters the tank from a supply via a control valve V_c. The control valve is adjustable, either manually or by some type of actuator. This may be an electric motor or a hydraulic pneumatic cylinder. Very often it would be a pneumatic diaphragm actuator. In general, increasing the pneumatic pressure above the diaphragm pushes it down against a spring and increases valve opening.

Open-loop control. In this form of control, the valve is adjusted to make output c equal to input r, but not readjusted continually to keep the two equal. Open-loop control, with certain safeguards added, is very common. For example, in the context of sequence control, that is, guiding a process through a sequence of predetermined steps. However, for systems such as the one at hand, this form of control will normally not yield high performance. A difference between input and output, a system error $e=r-c$ would be expected to develop, due to two major effects:

1. Disturbances acting on the system;
2. Parameter variations of the system.

These are prime motivations for the use of feedback control. For the water level example, pressure variations upstream of V_c and downstream of V_o can be important disturbances affecting inflow and outflow, and hence the level. A sudden or gradual change of flow resistance of the valves due to foreign matter or valve deposits represents a system parameter variation.[1] In a broader context, not only are the values of the parameters of a process often not precisely known, but they may also change greatly with operating condition.

Closed-loop control or feedback control. To improve performance, the operator could continuously readjust the valve based on observation of the system error e. A feedback control system in effect automates this action, as follows:

The output c is measured continuously and fed back to be compared with the input r. The error $e=r-c$ is used to adjust the control valve by means of an actuator.

The feedback loop causes the system to take corrective action if output c (actual level) deviates from input r (desired level), whatever the reason.

A broad class of systems can be represented by the block diagram shown in Fig.14.1. The sensor in Fig.14.1 measures the output c, and depending on its type, represents it by an electrical, pneumatic, or mechanical signal. The input r is represented by a signal in the same form. The summing junction or error junction is a device that combines the inputs to it according to the signs associated with the arrows: $e=r-c$.

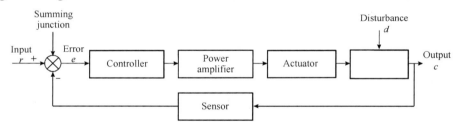

Fig.14.1　System block diagram

It is important to recognize that if the control system is good enough, the error e will usually be small, ideally zero. Therefore, it is quite inadequate to operate an actuator. A task of the controller is to amplify the error signal. The power amplifier raises power to the levels needed for the actuator.

14.3　Applications of Automatic Control

Although the scope of automatic control is virtually unlimited, we will limit this discussion to examples which are commonplace in modern industry.

Servomechanisms. Although a servomechanism is not a control application *per se*, this device is commonplace in automatic control. A servomechanism, or "servo" for short, is a closed-loop control system in which the controlled variable is mechanical position or motion.

It is designed so that the output will quickly and precisely respond to a change in the input command. Thus we may think of a servomechanism as a following device.

Another form of servomechanism in which the rate of change or velocity of the output is controlled is known as a rate or velocity servomechanism.

Process control. Process control is a term applied to the control of variables in a manufacturing process. Chemical plants, oil refineries, food processing plants, blast furnaces, and steel mill are examples of production processes to which automatic control is applied. Process control is concerned with maintaining at desired value such process variables as temperature, pressure, flow rate, liquid level, viscosity, density, and composition.

Much current work in process control involves extending the use of the digital computer to provide direct digital control (DDC) of the process variables. In direct digital control, the computer calculates the values of the manipulated variables directly from the values of the set points and the measurements of the process variables.[2] The decisions of the computer are applied to digital actuators in the process. Since the computer duplicates the analog controller action, these conventional controllers are no longer needed.

Power generation. The electric power industry is primarily concerned with energy conversion and distribution. Large modern power plants which may exceed several hundred megawatts of generation require complex control systems to account for the interrelationship of many variables and provide optimum power production.[3] Control of power generation may be generally regarded as an application of process control, and it is common to have as many as 100 manipulated variables under computer control.

Automatic control has also been extensively applied to the distribution of electric power. Power systems are commonly made up of a number of generating plants. As load requirements fluctuate, the generation and transmission of power is controlled to achieve minimum cost of system operation. In addition, most large power systems are interconnected with each other, and the flow of power between systems is controlled.

Numerical control. There are many manufacturing operations such as boring, drilling, milling, and welding which must be performed with high accuracy on a repetitive basis. Numerical control (NC) is a system that uses predetermined instructions called a program to control a sequence of such operations. These instructions are usually stored in the form of numbers, hence the name numerical control. The instructions identify what tool is to be used, in what way (e.g., cutting speed), and the path of the tool movement (position, direction, velocity, etc.).

Transportation. To provide mass transportation systems for modern urban areas, large complex control systems are needed. Several automatic transportation systems now in operation have high-speed trains running at several-minute intervals. Automatic control is necessary to maintain a constant flow of trains and to provide comfortable acceleration and braking at station stops.

Aircraft flight control is another important application in the transportation field. This has been proven to be one of the most complex control applications due to the wide range of system parameters and the interaction between controls. Aircraft control systems are frequently adaptive in nature, that is, the operation adapts itself to the surrounding conditions. For examples, since the behavior of an aircraft may differ radically at low and high altitudes the control system must be modified as a function of altitude.

Ship steering and roll-stabilization controls are similar to flight control but generally require far higher powers and involve lower speeds of response.

14.4 Artificial Intelligence in Mechatronics

Recent work using artificial intelligence has attempted to integrate various process control modules to increase productivity and quality in manufacturing operations. Further, the developments in expert systems, fuzzy logic, and neural networks are expected to be used at the higher level in the control hierarchy for machining processes.

Artificial neural networks in mechatronics. A recent trend in the area of automated manufacturing is the incorporation of artificial intelligence to enhance the on-line process control and inspection. Intelligent on-line process control and inspection in modern manufacturing systems have significant potential for improving production performance and product quality. Various studies suggest that incorporation of artificial neural networks results in a very promising synergy in intelligent manufacturing systems and processes. One reason for this is the dependence on knowledge-based systems, which are widely used and acknowledged for their relevancy, consistency, organization, and completeness.[4] The problem of acquiring expert knowledge in a form usable by an expert system is known as knowledge acquisition. This has been identified as a major bottleneck to the implementation of knowledge-based system technology. However, the acquisition process ends with the implementation of a knowledge-based system that cannot adapt to change and is unable to handle situations slightly different from known prototype conditions.[5] This has paved the way for the development of brain-like computation, namely artificial neural networks that can efficiently conduct production process control and inspection with a wide range of tolerance and uncertainties.[6]

The field of artificial intelligence (AI), particularly the technologies of artificial neural networks, is very useful at higher levels in the control hierarchy of manufacturing processes. The two basic AI approaches for providing decision support have been the macro, or top-down, modeling of human intelligence and the micro, or bottom-up, modeling of human intelligence.[7]

1. In the macro approach, as in knowledge-based systems, the human brain is treated as a black-box, and the human reasoning process is modeled through cognitive analysis of the decision-making tasks as described by the expert.
2. In the micro approach, as in neural networks, the human brain is treated as a white-box, and the human reasoning process is modeled through the observation of neural connections in the brain.

On the most fundamental level, neural networks perform pattern recognition more effectively than any other technique known. Once the network has detected a pattern, the information can be used to classify, predict, and analyze the causes of the pattern. Below are some unusual features of neural networks.

1. Recall of information is highly resistant to hardware failure (in contrast to traditional computers, in which retrieval can fail as result of a single memory element).
2. Positive and negative aspects of information are automatically balanced as a network reorganizes to solve a problem.
3. Abstraction of data occurs automatically as a by-product of learning information.
4. Pattern recognition occurs via parallel consideration of multiple constraints.
5. Hierarchical data structures can be conveniently represented as multiple-layer networks.
6. Networks can exhibit properties reminiscent of adaptive biological learning and can select and generate their own pattern features from exposure to stimuli.[8]
7. ANN can capture patterns occurring both in time (for example, auditory information) and space (such as visual data) and can operate in discrete or continuous representation modes.

From an engineering perspective, this new paradigm is a very powerful method for searching through solution space; however, its wide-range applications also include other fields such as business, medical, etc. ANN is best at solving problems that involve pattern recognition, adaptation, generalization, and prediction. Its implementation also offers particular advantages when solving problems that are very noisy, in which the performance of the system cannot be measured with each example, or potentially small improvements in performance can result in a substantial advantage in resource allocation and profits.[9] In the context of diagnostics, ANN is most valuable in applications that process continuous inputs, such as signal data. ANN solutions are working hard at detecting fraud (credit applications, insurance and warranty claims, and credit card fraud), in modeling and forecasting (bankruptcy prediction, credit scoring, securities trading, portfolio evaluation, mail list management, production marketing, and targeted marketing), and in process management (process modeling, process control, oil and gas exploration, reservoir management, production line management, machine diagnostics, flaw detection, product development, and industrial inspection). However, for ANN to reach its maximum potential, supporting hardware is needed that will make the network faster and thus more practical.

Knowledge-based System. Knowledge-based systems are the part of the AI field that focuses mainly on replicating cognitive human behavior. They do so by capturing problem-solving expertise of experts within a narrow problem domain and making it available for other organizational systems. The expertise is typically stored in a knowledge base in static form of If-Then-Else rules or a hierarchy of frames and objects. The knowledge base is used by an inference engine, which reasons with the knowledge in a serial manner, and applied to the different problem presented by user. Conventional knowledge-based systems are generally static, which make them inflexible in dynamic environments that require constant learning based on the action taken and feedback obtained from the action. Thus, an intelligent process control and inspection system cannot rely solely on a conventional knowledge-based system technology that uses a static knowledge base. Rather, it needs a dynamic learning mechanism that can help the system to deal with an uncertain reasoning environment with high flexibility and adaptability to the change in its condition.

Knowledge acquisition is probably the most difficult step in the development of an expert or knowledge-based system. The complexity of the problem at hand and the human factors of interacting and understanding the decision-making process of individuals combine to make the task of knowledge acquisition one that requires special tools, time, and considerable skill to perform accurately and effectively.[10] Difficulties with the expert may arise if the information provided is incorrect or is misinterpreted by the knowledge engineer. In addition, although experts may be highly skilled at solving the problem, they may be limited in their ability to describe the decision-making process in the meticulous detail required to make the expert system function properly. The expert may also provide extraneous information that can be eliminated from the knowledge base without affecting the decision-making capability.

Notes

[1] A sudden or gradual change of flow resistance of the valves due to foreign matter or valve deposits represents a system parameter variation.

句意：由于外来因素或阀门沉渣所引起的阀门流阻的突然的或逐渐的变化，代表了一个系统的参数变化。

[2] In direct digital control, the computer calculates the values of the manipulated variables directly from the values of the set points and the measurements of the process variables.

句意：在直接数字控制中，计算机直接根据设定点的数值和过程变量的测量值计算出被控制量的值。

[3] Large modern power plants which may exceed several hundred megawatts of generation require complex control systems to account for the interrelationship of the many variables and provide optimum power production.

句意：发电量可能超过几十万千瓦的现代化大型电厂需要复杂的控制系统来说明多个变量之间的相互关系，并提供最佳的发电量。

[4] One reason for this is the dependence on knowledge-based systems, which are widely used and acknowledged for their relevancy, consistency, organization, and completeness.

句意：其理由是可以可靠地利用这种基于知识的系统，这样的系统使用较广，主要是由于它们具有很好的关联性、协同性、组织性和完整性。

[5] However, the acquisition process ends with the implementation of a knowledge-based system that cannot adapt to change and is unable to handle situations slightly different from known prototype conditions.

句意：然而，获取过程以执行基于知识的系统结束，这个系统不能适应变化，不能处理相对于已知原型条件有轻微变化的情况。

[6] This has paved the way for the development of brain-like computation, namely artificial neural networks that can efficiently conduct production process control and inspection with a wide range of tolerance and uncertainties.

句意：这就为类脑计算的发展铺平了道路，这种计算被称为人工神经网络。人工神经网络可以有效处理具有大范围允许偏差和不确定性的生产过程控制和检测。

[7] The two basic AI approaches for providing decision support have been the macro, or top-down, modeling of human intelligence and the micro, or bottom-up, modeling of human intelligence.

句意：人工智能可以通过两种手段提供决策支持：一种是宏观的，即由上而下的人工智能建模；另一种是微观的，也就是自下而上的人工智能建模方式。

[8] Networks can exhibit properties reminiscent of adaptive biological learning and can select and generate their own pattern features from exposure to stimuli.

句意：网络显示的特性可以使人联想到自适应生物学习，能够从受到的刺激中选择并产生自己的模型特性。

[9] Its implementation also offers particular advantages when solving problems that are very noisy, in which the performance of the system cannot be measured with each example, or

potentially small improvements in performance can result in a substantial advantage in resource allocation and profits.

句意：当解决含有各种干扰因素的问题时，它的执行也提供了特别的优势，在这类问题中，系统的性能不能通过一个一个的例子来检测，或者可能在性能上很小的改善都会对资源分配和收益有很大的好处。

[10] The complexity of the problem at hand and the human factors of interacting and understanding the decision-making process of individuals combine to make the task of knowledge acquisition one that requires special tools, time, and considerable skill to perform accurately and effectively.

句意：所遇问题的复杂性及互相影响与理解个体决策过程中人的因素结合起来，使得知识获取的任务需要专门工具、时间和相当多的技能，这样才能准确、有效地执行。

Glossary

actuator   *n.* 启动器，执行器
adaptive   *adj.* （自）适应的
allocation   *n.* 配置
analog controller   模拟控制器
artificial intelligence   人工智能
artificial neural network   人工神经网络
bankruptcy   *n.* 破产，倒闭
boring   *n.* 镗孔
bottleneck   *n.* 瓶颈
by-product   *n.* 副产品
cistern   *n.* 水箱，水池，贮水器
closed-loop   *n.* 闭合回路，闭环
closed-loop control system   闭环控制系统
cognitive   *adj.* 认知的，认识的
commonplace   *adj.* 平常的   *n.* 平凡常见的事
completeness   *n.* 完整性
consistency   *n.* 一致性
cylinder   *n.* 汽缸
decision-making   *n.* 决策
diaphragm   *n.* 隔膜，膜片

digital actuator 数字执行器
direct digital control (DDC) 直接数字控制
disturbance torque 干扰力矩
duplicate *v.* 加倍，复制
expertise *n.* 专家技能
exposure to sth. 暴露于……，受到……
extraneous *adj.* 外来的，无关的
flow rate 流量
fluctuate *v.* 波动，起伏
following device 随动装置
fraud *n.* 欺骗，欺诈
hierarchy *n.* 分层结构
hydraulic *adj.* 液压的，水压的
incorporation *n.* 结合，合并
insurance and warranty claim 保险与担保索赔
liquid level 液位
manipulated variable 操纵变量
megawatt *n.* 兆瓦，百万瓦
meticulous *adj.* 细心的，小心翼翼的
milling *n.* 铣削
optimum *adj.* 最佳的，最优的
pneumatic *adj.* 空气的，气动的
predetermine *v.* 预定，先定
promising *adj.* 有希望的
prototype *n.* 原型
portfolio *n.* 投资组合
punched paper tape 穿孔纸带
radically *adv.* 根本上
reasoning *n.* 推论，推理
relevancy *n.* 关联性
reminiscent of 发人回想的，发人联想的
safeguard *n.* 保护措施，防护设施
servomechanism *n.* 伺服机构，伺服机械
ship-steering 船舶转向

steering　*n.* 驾驶，转向，操舵
substantial　*adj.* 实质上的，相当的，大量的
summing junction　求和节点
synergy　*n.* 协同作用
tolerance　*n.* 宽容，容忍，容许量
viscosity　*n.* 黏度，黏性

Unit 15   CAD/CAM

15.1   Introduction

Computer-aided design (CAD) involves the use of computers to create design drawings and product models. Compute-aided design is usually associated with interactive computer graphics, known as a CAD system. CAD systems are powerful tools and are used in the design and geometric modeling of components and products.

Drawings are generated at workstations, and the design is displayed continuously on the monitor in different colors for its various components. The designer can easily conceptualize the part to be designed on the graphics screen and can consider alternative designs or modify a particular design to meet specific design requirements. Using powerful software such as CATIA (computer-aided three-dimensional interactive applications), the design can be subjected to engineering analysis and can identify potential problems, such as an excessive load, deflection, or interference at mating surfaces during assembly.[1] Information (such as list of materials, specifications, and manufacturing instructions) can also be stored in the CAD database. Using this information, the designer can analyze the manufacturing economics of alternative designs.

15.2   Geometric Modeling

In geometric modeling, a physical object or any of its parts is described mathematically or analytically. The designer first constructs a geometric model by giving commands that create or modify lines, surfaces, solids, dimension and text. Together, these present an accurate and complete two- or three-dimensional representation of the object. The results of these commands are displayed and can be moved around on the screen, and any section desired can be magnified to view details. These data are stored in the database contained in computer memory.

The models in a CAD system can be presented in three different ways:

1. In line representation [wire-frame model shown in Fig.15.1(a)], all of the edges of the model are visible as solid lines. However, this image can be ambiguous, particularly for complex shapes. Therefore, various colors generally are used for different parts of the object to make it easier to visualize.

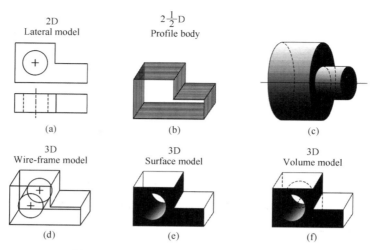

Fig.15.1 Various types of modeling for CAD

The three types of wire-frame representations are two, two and one-half, and three dimensional. A two-dimensional image shows the profile of the object. A two and one-half dimensional image can be obtained by translational sweep, that is, by moving the two-dimensional object along the 2 axes. For round objects, a two and one-half dimensional model can be generated by simply rotating a two-dimensional model around its axis.[2]

2. In the surface model, all visible surfaces are shown in the model. Surface models define surface features and edges of objects. Modern CAD programs use Bezier curves, B-splines, or non-uniform rational B-splines (NURBS) for surface modeling. Each of these uses control points to define a polynomial curve or surface.[3]

The drawback to Bezier curves is that the modification of one control point will affect the entire curve. B-splines are a blended piecewise polynomial curve, where modification of a control point only affects the curve in the area of the modification.

3. In the solid model, all surfaces are shown, but the data describe the interior volume. Solid models can be constructed from "swept volumes" (Fig.15.1) or by the techniques shown in Fig.15.2. In boundary representation (B-rep), surfaces are combined to develop a solid model (Fig.15.2). In constructive solid geometry (CSG), simple shapes such as spheres, cubes, blocks, cylinders, and cones (called primitives of solids) are combined to develop a solid model.

Programs are available whereby the user selects any combination of these primitives and their sizes and combines them into the desired solid model. Although solid models have certain advantages (such as ease of design analysis and ease of preparation for manufacturing the part), they require more computer memory and processing time than the wire-frame and surface models shown in Fig.15.1.

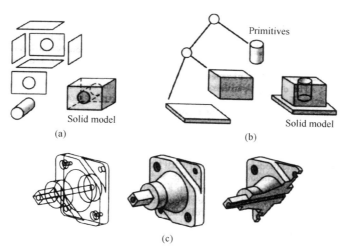

Fig.15.2 (a) Boundary representation of solids; (b) A solid model composed of solid primitives; (c) Three representations of the same part

15.3 CAD/CAM

There is no need to give unnecessary details about the importance of product quality. Quality must be built into the product, but high quality does not necessarily mean higher cost, and marketing poor-quality products indeed can be very costly to the manufacturer. It has been shown that high quality is far more attainable and less expensive if design and manufacturing activities are integrated properly rather than if they were treated as separate entities.[4] Integration can be performed successfully and effectively through computer-aided design, engineering, manufacturing, process planning, and the simulation of processes and systems. The widespread availability of high-speed processors and highly developed software has allowed computers to proliferate into all areas of manufacturing. Computers are used to perform a wide variety of tasks ranging from individual part drafting, to modeling manufacturing processes and systems, and to performing database management.

Manufacturing is a complex system, because it consists of many diverse physical as well as human elements, some of which are difficult to predict and control. These difficulties include such factors as the supply and cost of raw materials, national and global market changes, the impact of constantly developing technologies, and human behavior and performance. Ideally, a manufacturing system should be represented by mathematical and physical models which show the nature and extent of the interdependence of all of the variables involved. In this way, the effects of a change or a disturbance that occurs anywhere in the system can be analyzed, and necessary adjustments can be made.

For example, the supply of a particular raw material may be reduced significantly due to global demands, wars, strikes, or geopolitics. Because the raw material cost now will rise, alternative materials may have to be considered and selected. This selection must be made after careful consideration of several factors because such a change may have adverse effects on product quality, production rate, and manufacturing costs. In a constantly changing marketplace, the demand for a product may fluctuate randomly and rapidly for a variety of reasons. As examples, note the downsizing of automobiles during the 1980s in response to fuel shortages and, in contrast, the recent popularity of sport-utility vehicles and current interest in gas-electric hybrid vehicles.[5] The system thus must be able to produce the modified product in a relatively short lead time and minimize large expenditures in machinery and tooling.

Such a complex system can be difficult to analyze and model because of a lack of comprehensive or reliable data on all of the variables involved. Furthermore, it is not easy to correctly predict and control some of these variables. For example: (1) machine-tool characteristics, their performance, and their response to random external disturbances cannot be precisely modeled; (2) raw-material costs are difficult to predict accurately; and (3) human behavior and performance are difficult to model. In spite of the difficulties, much progress has been made in modeling and simulating manufacturing systems.

Design analysis and optimization. After the geometric features of a particular design have been determined, the design is subjected to an engineering analysis. This phase may consist of analyzing stresses, strains, deflections, vibrations, heat transfer, temperature distribution, or dimensional tolerances. Various sophisticated software packages are available, each having the capabilities to compute these quantities accurately and rapidly.

Because of the relative ease with which such analyses can now be done, designers are increasingly willing to analyze a design more thoroughly before it is put into production. Experiments and measurements in the field may be necessary to determine the actual effects of loads, temperature, and other variables on the designed components.

Design review and evaluation. An important design stage is the design review and evaluation used to check for any interference between various components. This is done in order to avoid difficulties during assembly or in use of the part and to determine whether moving members (such as linkages) are going to operate as intended. Software is available with animation capabilities to identify potential problems with moving members and other dynamic situations. During the design review and evaluation stage, the part is dimensioned and toleranced precisely to the full degree required for manufacturing it.

Documentation and drafting. After the preceding stages have been completed, the design is reproduced by automated drafting machines for documentation and reference. At this stage, detailed and working drawings also are developed and printed.

Database. Many components either are standard components that are mass produced according to a given design specification (such as bolts or gears) or are identical to the parts used in previous designs. Modern CAD systems thus have a built-in database management system that allows designers to locate, view, and adopt parts from a stock part library. These parts can be modeled parametrically to allow cost-effective updating of the geometry. Some databases are available commercially with extensive parts libraries; many vendors make their part libraries available on the world-wide web.

Computer-aided manufacturing (CAM) involves the use of computers to assist in all phases of manufacturing a product. Because of the joint benefits, computer-aided design and compute-aided manufacturing often are combined into CAD/CAM systems. This combination allows the transfer of information from the design stage into the stage of planning for manufacturing without the need to reenter the data of part geometry manually. The database developed during CAD is stored and processed further by CAM into the necessary data and instructions for operating and controlling production machinery, material-handling equipment, and automated testing and inspection for product quality. CAD/CAM systems also are capable of coding and classifying parts into groups that have similar shapes using alphanumeric coding.[6]

An important feature of CAD/CAM in machining operations is its capability to describe the tool path. The instructions (programs) are computer generated, and they can be modified by the programmer to optimize the tool path. The engineer or technician can display and check the tool path for possible tool collisions with clamps, fixtures or other interferences.

By standardizing product development and reducing design effort, tryout, and prototype work, CAD/CAM has made possible significantly reduced manufacturing costs and improved productivity. For example, the two-engine Boeing 777 passenger airplane was designed completely by computer (paperless design) with 2000 workstations linked to eight computers. The plane was constructed directly from the CAD/CAM software developed (an enhanced CATIA system), and no prototypes or mockups were built, as were required for previous models.[7] The cost for this development was on the order of $ 6 billion.

Some typical applications of CAD/CAM are as follows:
1. Programming for numerical control and industrial robots;
2. Design of dies and molds for casting, for example, in which shrinkage allowances are preprogrammed;
3. Dies for metalworking operations, such as complex dies for sheet forming and progressive dies for stamping;
4. Design of tooling and fixtures and EDM electrodes;
5. Quality control and inspection, such as coordinate-measuring machines (CMM) programmed on a CAD/CAM workstation;
6. Process planning and scheduling;
7. Plant layout.

15.4 Computer-aided Process Planning

Process planning represents the link between engineering design and shop-floor manufacturing. It is concerned with selecting methods of production: tooling, fixtures, machinery, sequence of operations, and assembly. All of these diverse activities must be planned, which traditionally has been done by process planners. The sequence of processes and operations to be performed, the machines to be used, the standard time for each operation, and similar information are all documented on a routing sheet.

When done manually, this task is highly labor-intensive and time-consuming and relies heavily on the experience of the process planner. Modern practice in routing sheets is to store the relevant data in computers and affix a bar code (or other identification) to the part. In order to shorten the gap between CAD and CAM, many computerized process planning approaches have been developed to accomplish the task of process planning in the past decades. These approaches are known as computer-aided process planning (CAPP). With the help of CAPP, better and faster process plans can be generated. CAPP can be classified into three categories: indexed CAPP, variant CAPP and generative CAPP. Each approach is appropriate under certain conditions.[8]

Notes

[1] Using powerful software such as CATIA (computer-aided three-dimensional interactive applications), the design can be subjected to engineering analysis and can identify potential problems, such as an excessive load, deflection, or interference at mating surfaces during assembly.

句意：利用功能强大的软件，如 CATIA（计算机辅助三维交互应用），对完成的设计进行工程分析并且可以找出潜在的问题，如过负荷、变形或者装配过程中发生配合面的干涉现象等。

[2] The three types of wire-frame representations are two, two and one-half, and three dimensional. A two-dimensional image shows the profile of the object. A two and one-half dimensional image can be obtained by translational sweep, that is, by moving the two-dimensional object along the 2 axes. For round objects, a two and one-half dimensional model can be generated by simply rotating a two-dimensional model around its axis.

句意：线框表达方式有 3 种，即二维、二维半和三维表达图。二维图可显示一个物件的轮廓；二维半图可以通过平动扫掠，即将一个二维图沿着它的两根轴平行移动就能获得。对于回转体，其二维半模型只要将一个二维图模型绕着某根轴旋转即可生成。

[3] In the surface model, all visible surfaces are shown in the model. Surface models define surface features and edges of objects. Modern CAD programs use Bezier curves, B-splines, or non-uniform rational B-splines (NURBS) for surface modeling. Each of these uses control points to define a polynomial curve or surface.

句意：在表面模型中，所有可见表面均可在该模型上显示出来。表面模型可以定义物体的表面和棱边的特征。现代 CAD 程序可采用 Bezier 曲线、B 样条曲线和非均匀有理 B 样条（NURBS）曲线进行表面建模。这 3 种曲线均可利用控制点来定义一个多项式曲线或曲面。

[4] It has been shown that high quality is far more attainable and less expensive if design and manufacturing activities are integrated properly rather than if they were treated as separate entities.

句意：事实已表明，如果将设计与制造活动很好地集成起来，而不是将其割裂开来单独处理，那么要获得优质的产品将会容易得多，而且成本也会降低。

[5] In a constantly changing marketplace the demand for a product may fluctuate randomly and rapidly for a variety of reasons. As examples, note the downsizing of automobiles during the 1980s in response to fuel shortages and, in contrast, the recent popularity of sport-utility vehicles and current interest in gas-electric hybrid vehicles.

句意：在当今不断变化的市场中，产品需求也会由于各种原因而随时发生迅速的变化。例如，我们注意到，20 世纪 80 年代为应付燃料短缺，汽车都做得很小，而现在又开始流行运动型车和油电混合车。

[6] The database developed during CAD is stored and processed further by CAM into the necessary data and instructions for operating and controlling production machinery, material-handling equipment, and automated testing and inspection for product quality. CAD/CAM systems also are capable of coding and classifying parts into groups that have similar shapes using alphanumeric coding.

句意：CAD 产生的数据库存储在 CAM 中，并被进一步处理成操作和控制生产机械、物料处理设备及产品质量的自动测试和检测所必需的数据和指令。CAD/CAM 系统还能对零件进行编码处理，用字母字符代码将形状类似的零件进行分组。

[7] For example, the two-engine Boeing 777 passenger airplane was designed completely by computer (paperless design) with 2000 workstations linked to eight computers. The plane was constructed directly from the CAD/CAM software developed (an enhanced CATIA system), and no prototypes or mockups were built, as were required for previous models.

句意：例如，双引擎波音 777 客机就是完全靠 2000 台联机的计算机工作站完成设计任务的。该客机直接采用所开发的 CAD/CAM 软件（一种改善的 CATIA 系统）构建，不像以前需要制作样机或 1∶1 大小的实体模型。

[8] In order to shorten the gap between CAD and CAM, many computerized process planning approaches have been developed to accomplish the task of process planning in the past decades. These approaches are known as computer-aided process planning (CAPP). With the help of CAPP, better and faster process plans can be generated. CAPP can be classified into three categories: indexed CAPP, variant CAPP and generative CAPP. Each approach is appropriate under certain conditions.

句意：为了将 CAD 和 CAM 连成一体，过去几十年里已开发出了许多采用计算机来进行工艺规划的手段，它们被称为计算机辅助工艺规划——CAPP。借助 CAPP，工艺规划可以做得更快更好。CAPP 可以分为检索式、样件式和创成式 3 种类型。它们都有各自的适用条件。

Glossary

Bezier curve   贝塞尔曲线

B-spline   B 样条（曲线、曲面）

conceptualize   *n.* 概念化，抽象化

constructive solid geometry   构造实体几何（表达方式）

cost-effective   划算的，具有成本效益的

computer-aided process planning (CAPP)   计算机辅助工艺规划

downsizing   *n.* 减小规模，缩小

gas-electric hybrid vehicle   油电混合车

geopolitics   *n.* 地缘政治关系

lead time   交货期，生产准备周期

mockup   *n.* 同实物等大的（研究用）模型

non-uniform rational B-spline   非均匀有理 B 样条（曲线、曲面）

optimization   *n.* 优化

polynomial   *n.* 多项式

polynomial curve   多项式曲线

primitive   *n.* 基本单元   *adj.* 原始的

proliferate   *v.* 普及，推广，扩展

shrinkage allowance   （凝固）收缩余量

spline   *n.* 样条函数[数]，花键[机]

tolerance   *n.* 公差   *v.* 规定公差范围

tool path   刀具（加工）路径

wire-frame representation   线框（图）表达

Unit 16 Transducers

16.1 Introduction

A transducer is a device which converts the quantity being measured into an optical, mechanical, or more commonly, electrical signal. The energy-conversion process is referred to as transduction.[1]

16.2 Transducer Elements

Although there is exception, most transducers consist of a sensing element and a conversion or control element, as shown in the two-block diagram of Fig.16.1.

Fig.16.1 Two-block diagram representation of a typical transducer

For example, diaphragms, bellows, strain tubes and rings, Bourdon tubes, and cantilevers are sensing elements which respond to changes in pressure or force and convert these physical quantities into displacements.[2] This displacement may then be used to change into an electrical parameter such as voltage, resistance, capacitance, or inductance. Such combinations of mechanical and electrical elements form electromechanical transducing devices or transducers. Similar combinations can be made for other energy input such as thermal, photo, magnetic and chemical, giving thermoelectric, photoelectric, electromagnetic, and electrochemical transducers respectively.

16.3 Analog and Digital Transducers

Considerable experience has been accumulated with analog transducers, signal conditioning, A/D converters, etc., and it is natural that the majority of current systems tend to use these techniques. However, there are a number of measuring techniques that are essentially digital in nature, and when used as separate measuring instruments they require some integral digital circuitry, such as frequency counters and timing circuits, to provide an indicator output.[3] This type of transducer, if coupled to a computer, does not necessarily require the same amount of equipment, since much of the processing done by the integral circuitry could be programmed and performed by the computer.

Collins classifies the signals handled in control and instrumentation systems as follows:

1. **Analog,** in which the parameter of the system to be measured although initially derived in an analog form by a sensor, is converted to an electrical analog, either voltage or current. Some averaging usually occurs, either by design or inherent in the methods adopted.

2. **Coded-digital,** in which a parallel digital is generated, each bit radix-weighted according to some predetermined code.[4] These are referred to as direct digital transducers.

3. **Digital,** in which a function, such as mean rate of a repetitive signal, is a measure of the parameter being measured. These are subsequently referred to as frequency-domain transducers.

Some analog transducers are particularly suited to conversion to digital outputs using special techniques. The most popular of these are output synchros, and similar devices which produce a modulated output of a carrier frequency. For ordinary analog use, this output has to be demodulated to provide a DC signal whose magnitude and sign represents any displacement of the transducer's moving element. Although it is then possible to use a conventional A/D technique to produce a digital output, there are techniques by which the synchro output can be converted directly to a digital output while providing a high accuracy and resolution, and at a faster rate than is possible in the A/D converter method.

Direct digital transducers are, in fact, few and far between, since there do not seem to be any natural phenomena in which some detectable characteristic changes in discrete intervals as a result of a change of pressure, or change of temperature etc.[5] There are many advantages in using direct digital transducers in ordinary instrumentation systems, even if computers are not used in the complete installation. These advantages are:

1. The ease of generating, manipulating and storing digital signals, as punched tape, magnetic tape etc;
2. The need for high measurement accuracy and discrimination (resolution);
3. The relative immunity of a high-level digital signal to external disturbances (noise);
4. Ergonomic advantages in simplified data presentation (e.g. digital readout avoids interpretation errors in reading scales or graphs);
5. Logistic advantage concerning maintenance and spares compared with analog or hybrid systems.

The most active development in direct digital transducers has been in shaft encoders, which are used extensively in machine tools and in aircraft systems. High resolution and accuracies can be obtained, and these devices may be mechanically coupled to provide a direct digital output of any parameter which gives rise to a measurable physical displacement. For example, a shaft encoder attached to the output shaft of a Bourdon tube gauge can be used for direct pressure measurement or temperature measurement using vapour pressure thermometers. The usual disadvantage of these systems is that the inertia of the instrument and encoder often limits the speed of response and therefore the operation frequencies.

Frequency domain transducers have a special part to play in on-line systems with only a few variables to be measured, since the computer can act as part of an A/D conversion system and use its own registers and clock for counting pulses or measuring pulse width.[6] In designing such systems, consideration must be given to the computer time required to access and process the transducer output data.

16.4 Use of Sensors in Programmable Automation

In this section we are concerned with the application of sensor-mediated programmable automation to material-handling, inspection, and assembly operations in batch-produced, discrete-part manufacturing.

Programmable automation consists of a system of multi-degree-of-freedom manipulators (commonly known as industrial robots) and sensors under computer control, which can be programmed to perform specified jobs in the manufacturing process and can be applied to new (but similar) jobs by reprogramming. This is particularly important where production runs are small and where different models may have to be produced frequently. Today industrial robots have contact sensors as aids to manipulation, non-contact sensors as aids to recognition, inspection, or manipulation of workpieces.

Extending the present capabilities of industrial robots will require a considerable improvement in their capacity to perceive and interact with the surrounding environment. In particular, it is desirable to develop sensor-mediated, computer-controlled interpretive systems that can emulate human capabilities. To be acceptable by industry, these hardware/software systems must perform as well or better than human workers. Specifically, they must be inexpensive, fast, reliable, and suitable for the factory environment.

Sensors can be broadly divided into three areas of application: visual inspection, finding parts, and controlling manipulation.

Visual inspection. Here we are concerned only with an important aspect of visual inspection: the qualitative and semiquantitative type of inspection performed by human vision rather than by measuring instruments. Such inspection of parts of assemblies includes identifying parts; detecting of burrs, cracks, and voids; examining cosmetic qualities and surface finish; counting the number of holes and determining their locations and sizes; assessing completeness of assembly; and so on. It is evident that a large library of computer programs will have to be developed to cope with the numerous classes of inspection.

Finding parts. For material-handling and assembly operations in the unstructured environment of the great majority of factories, it will probably be necessary to "find" workpieces—that is, to determine their positions and orientations and sometimes also to identify them.[7] Thus it is necessary to augment existing robots with visual sensors to be able to determine the identity, position, and orientation of parts and to perform visual inspection.

Controlling manipulation. It appears useful to consider the use of both contact and noncontact sensors in manipulator control and to try to assess where each sensor is most appropriate. One approach is to divide the sensory domain into coarse and fine sensing, using non-contact sensors for coarse resolution and contact sensors for fine resolution. For example, in acquiring a workpiece that may be randomly positioned and oriented, a visual sensor may be used to determine the relative position and orientation of the workpiece rather coarsely, say, to one tenth of an inch. From this information the manipulator can be positioned automatically. The somewhat compliant fingers of the manipulator hand, bracketing the workpiece, will now be close enough to effect closure, relying on touch sensors to stop the motion of each finger when a specified contact pressure is detected.[8] After contacting the workpiece without moving it, the compliant fingers have flexed no more than a few thousandths of an inch before a stopping. The touch sensors have thus performed fine resolution sensing and have compensated for the lack of precision of both the visual sensor and the manipulator.

Other common applications for contact sensors, which entail fine resolution or precision sensing, include:

1. Collision avoidance, using force sensors on the links and hand of a manipulator. Motion is quickly stopped when any one of preset force thresholds is exceeded.
2. Packing operations, in which parts are packed in orderly fashion in tote boxes. Force sensors can be used to stop the manipulator when its compliantly mounted hand touches the bottom of the box, its sides, or neighboring parts. This mode of force feedback compensates for the variability of the positions of the box and the parts and for the small but important variability of the manipulator positioning.
3. Insertions of pegs, shafts, screws and bolts into holes. Force and torque sensors can provide feedback information to correct the error of a computer-controlled manipulator.

16.5　Some Terms

Sensitivity. The relationship between the measurand and the transducer output signal is usually obtained by calibration tests and is referred to as the transducer sensitivity. Sensitivity is defined as the ratio of the change in output to the corresponding change in input under static or steady-state conditions.

Repeatability. The consistence among several repeatedly measured results for the same measurand under the same measuring condition.

Resolution. Two other terms associated with the quality of a measurement are precision and resolution. Resolution is defined as the minimum discernible (detectable) change in the measurement and that can be detected. The resolution of a measurement is not a constant for a given instrument but may be changed by the measurand or the test conditions. For example, a nonlinear meter scale has a higher resolution at one end than at the other due to the spacing of the scale divisions. Likewise, noise induced in a system can affect the ability to resolve a very small change in voltage or resistance. Temperature changes can also affect measurements because of the effect on resistance, capacitance, dimensions of mechanical part, drift, and so forth.

Precision. Precision is a measure of the repeatability of a series of data points taken in the measurement of some quantity. It means the consistence among various individual measured results. The precision of an instrument depends on both its resolution and its stability.

Stability refers to freedom from random variations in the result. A precise measurement requires both stability and high resolution. Precision is a measure of the dispersion of a set of data, not a measure of the accuracy of the data. It is possible to have a precision instrument that provides readings that are not scattered but that are not accurate because of a systematic error. However, it is not possible to have an accurate instrument unless it is also precise.

Accuracy. Data measured with test equipment are not perfect, rather, the accuracy of the data depends on the precision of the test equipment and the conditions under which the measurement was made. In order to interpret data from an experiment, we need to have an understanding of the nature of errors. Experimental error should not be thought of as a mistake. All measurements that do not involve counting are approximations of the true value. Error is the difference between the true or best accepted value of some quantity and the measured value. A measurement is said to be accurate if the error is small. Accuracy refers to the degree to which the measured result is close to its true value. It is important for the user of an instrument to know what confidence can be placed in it. Instrument manufacturers generally quote accuracy specifications in their literature, but the user needs to be cautioned to understand the specific conditions for which an accuracy figure is stated. The number of digits used to describe a measured quantity is not always representative of the true accuracy of the measurement.

Notes

[1] A transducer is a device which converts the quantity being measured into an optical, mechanical, or more commonly, electrical signal. The energy-conversion process is referred to as transduction.

句意：传感器是一种将被测量转换为光的、机械的或者更平常的电信号的装置。能量转换的过程称为换能。

[2] For example, diaphragms, bellows, strain tubes and rings, Bourdon tubes, and cantilevers are sensing elements which respond to changes in pressure or force and convert these physical quantities into displacements.

句意：如振动膜、波纹管、应力管和应力环、布尔登管（弹簧管）和悬臂梁都是敏感元件，它们对压力和力做出响应，并将这些物理量转变成位移。

[3] However, there are a number of measuring techniques that are essentially digital in nature, and which when used as separate measuring instruments they require some integral digital circuitry, such as frequency counters and timing circuits, to provide an indicator output.

句意：然而，很多测试技术本质上就是数字化的，当作为独立的测试仪器使用时，它们需要通过由数字积分电路组成的频率计数器和定时电路来提供输出指示。

[4] Coded-digital, in which a parallel digital is generated, each bit radix-weighted according to some predetermined code.

句意：在编码-数字式信号处理中，根据预定的编码方式，每一位按基数进行加权，由此产生一个并行数字。

[5] Direct digital transducers are, in fact, few and far between, since there do not seem to be any natural phenomena in which some detectable characteristic changes in discrete intervals as a result of a change of pressure, or change of temperature etc.

句意：直接数字式传感器实际上是极少的，因为似乎不存在某些可检测特性会随着压力或温度等的变化而发生不连续变化的任何自然现象。

[6] Frequency domain transducers have a special part to play in on-line systems with only a few variables to be measured, since the computer can act as part of an A/D conversion system and use its own registers and clock for counting pulses or measuring pulse width.

句意：频率式传感器在只有几个变量要测量的在线系统中具有特殊作用，因为计算机可作为 A/D 转换系统的一部分，并利用其自身的寄存器和时钟来计算脉冲数或测量脉冲宽度。

[7] For material-handling and assembly operations in the unstructured environment of the great majority of factories, it will probably be necessary to "find" workpieces—that is, to determine their positions and orientations and sometimes also to identify them.

句意：对于大多数工厂中非结构化环境下的物料装卸和装配操作，常常需要"寻找"工件，即确定工件的位置和方向，有时还要进行识别。

[8] The somewhat compliant fingers of the manipulator hand, bracketing the workpiece,

will now be close enough to effect closure, relying on touch sensors to stop the motion of each finger when a specified contact pressure is detected.

句意：操作臂的柔性手指，将会在很靠近工件时完成关闭动作，从而抓起工件，这些动作是依靠接触传感器完成的，当接触传感器检测到特定的接触压力时，就使每个手指的运动停止。

* transducer 和 sensor 的区别

在英文文献中常常遇到 transducer 和 sensor 这两个词混用。严格地说，transducer 称为换能器，它是将输入能量转换成另一种形式的能量的装置。而且，它在不同专业领域还有变送器（如闭环反馈装置中）和变频器（如电梯与空调的调速变频控制中）等名称。而 sensor 称为传感器、感测器，它将物理、化学、机械、光学等过程中可测量量转换成对技术人员或计算机有意义的数据。一般情况下，transducer 需要转换（放大）电路，sensor 有可能不需要这样的电路。例如，压电晶体、光电管属于 transducer，而普通温度计应该算是 sensor。还有一个词 probe，俗称测头、探测器、探头，一般是指简单的传感器，或 transducer 及 sensor 的口语或简单表达词；而 sensing element 是敏感元件如光敏电阻、热敏电阻等。

Glossary

accuracy   *n.* （数据，加工）精度
appropriate   *adj.* 合适的，恰当的
approximation   *n.* 大致估计，概算；近似
assembly   *n.* 装配，集会，集合
cantilever   *n.* 悬臂，悬臂梁
carrier   *n.* 载波（信号）
collision   *n.* 碰撞，冲突
compensate   *v.* 补偿，偿还
compliantly   *adv.* 顺从地
cosmetic   *adj.* 化妆用的；装点门面的，表面的
demodulate   *v.* 解调，检波
discernible   *adj.* 可辨别的，可识别的
ergonomic   *adj.* 人类工程学的
gauge   *n.* 标准规格，（金属的）厚；计量器；容量，限度，范围；域
hybrid   *n.* 混合源物，混合物
installation   *n.* 装置，设备；安装，装配
interpretive   *adj.* 解释的，翻译的，说明的

magnitude　*n.* 大小，积，量，长（度）；尺寸，幅度
maintenance　*n.* 维修，保养
manipulator　*n.* 操作的人；操纵者；机械手
measurand　*n.* 被测量值
orientation　*n.* 定位（向），取向；方向，方位
output synchros　同步输出器
peg　*n.* 销子
precision　*n.*（仪器，设备）精确性，精确度
resolution　*n.* 分辨率
scattered　*adj.* 稀疏的，分散的
sensitivity　*n.* 灵敏度，敏感性
synchro　*n.*（自动）同步机
threshold　*n.* 门槛；界限，限度；阈值

Unit 17 Robots

17.1 Introduction

In the early 1960s, the industrial revolution put industrial robots in the factory to release the human operator from risky and harmful tasks. The later incorporation of industrial robots into other types of production processes added new requirements that called for more flexibility and intelligence in industrial robots. Currently, the creation of new needs and markets outside the traditional manufacturing robotic market (i.e., cleaning, mine sweeping, construction, shipbuilding, agriculture) and the aging world we live in is demanding field and service robots to attend to the new market and human social needs.

17.2 Definition of Robot

The definition developed by the Robot Institute of America is given by the following description:

A robot is a reprogrammable multifunctional manipulator designed to move material, parts, tools, or specialized devices through variable programmed motions for the performance of a variety of tasks.[1]

The key words are reprogrammable and multifunctional, since most single-purpose machines do not meet these two requirements.

Reprogrammable means that the machine must be capable of being reprogrammed to perform a new or different task or to be able to change significantly the motion of the arm or tooling. Multifunctional emphasizes the fact that a robot must be able to perform many different functions, depending on the program and tooling currently in use.

Despite the tremendous capability of currently available robots, even the most poorly prepared worker is better equipped than a robot to handle many of the situations which occur in the work cell. Workers, for example, realize when they have dropped a part on the floor or when a parts feeder is empty. Without a host of sensors, a robot simply does not have any of his information; and even with the most sophisticated sensor system available, a robot cannot match an experienced human operator. The design of a good automated work cell therefore requires the use of peripheral equipment interfaced to the robot controller to even imitate the sensory capability of a human operator.

17.3 Components of a Robot System

The components of a robot system could be discussed either from a physical point of view or from a system point of view. Physically, we would divide the system into the robot, power system, and controller (computer). Likewise, the robot itself could be partitioned anthropomorphically into base, shoulder, elbow, wrist, gripper, and tool.

Consequently, we will describe the components of a robot system from the point of view of information transfer.[2] That is, what information or signal enters the component; what logical or arithmetic operation does the component perform; and what information or signal does the component produce? It is important to note that the same physical component may perform many different information processing operations (e.g., a central computer performs many different calculations on different data). Likewise, two physically separate components may perform identical information operations (e.g., the shoulder and elbow actuators both convert signals to motion in very similar ways).

Actuator. Associated with each joint on the robot is an actuator which causes that joint to move. Typical actuators are electric motors and hydraulic cylinders. Typically, a robot system will contain six actuators, since six degrees of freedom are required for full control of position and orientation. Many robot applications do not require this full flexibility, and consequently, robots are often built with five or fewer actuators.

Sensor. To control an actuator, the computer must have information regarding the position and possibly the velocity of the actuator. In this context, the term position refers to a displacement from some arbitrary zero reference point for that actuator. For example, in the case of a rotary actuator, "position" would really be the angular position and be measured in radians.

Many types of sensors can provide indications of position and velocity. The various types of sensors require different mechanism for interfacing to the computer. In addition, the industrial use of the manipulator requires that the interface be protected from the harsh electrical environment of factory. Sources of electrical noise such as arc welders and large motors can easily make a digital system useless unless care is taken in design and construction of the interface.

Computation. We could easily have labeled the computation module "computer", as most of the functions such as servo, kinematics, dynamics and workplace sensor analysis are typically performed by digital computers.[3] However, many of the functions may be performed in dedicated custom hardware or networks of computer. We will, thus, discuss the computational component as if it were a simple computer, recognizing that the need for real-time control may require special equipment and that some of this equipment may even be analog, although the current trend is toward fully digital systems.

In addition to these easily identified components, there are also supervisory operations such as path planning and operator interaction.

17.4 Industrial Robots

In Japan, the Japanese Industrial Robot Association (JIRA) classifies industrial robots by the method of input information and the method of teaching:

1. **Manual manipulators.** Manipulators directly activated by the operator.
2. **Fixed-sequence robot.** Robot that once programmed for a given sequence of operations is not easily changed.
3. **Variable-sequence robot.** Robot that can be programmed for a given sequence of operations and can easily be changed or reprogrammed.[4]
4. **Playback robot.** Robot that "memorizes" work sequences taught by a human being who physically leads the device through the intended work pattern; the robot can then create this sequence repetitively from memory.
5. **Numerically controlled (NC) robot.** Robot that operates from and is controlled by digital data, as in the form of punched tape, cards, or digital switches; operates like an NC machine.
6. **Intelligent robot.** Robot that uses sensory perception to evaluate its environment and make decisions and proceeds to operate accordingly.[5]

Current and emerging robot applications in industry can be categorized on the complexity and requirements of the job. They range from simple, low technology such as pick-and-place operations through medium technology, painting, some assembly and welding operations to high technology precision assembly and inspection operations.

Pick-and-place operations. The earliest applications of robots were in machine loading-unloading, pick-and-place, and material transfer operations. Such robots typically were not servo controlled and worked with pneumatic or hydraulic power. The load-carrying requirements were high, working in dirty or hazardous factory environments. Replacing unskilled human labor often in hazardous jobs, these robots had to be robust and low in initial and maintenance costs.[6]

Painting and welding operations. The next level in the sophistication of industrial robot applications was in spray painting, and spot and arc welding (see Fig.17.1). These applications complemented or replaced certain skilled human labor. Often the justification was by eliminating dangerous environmental exposures. These applications often require tracking complex trajectories such as painting surface contours, hence servo controlled articulated or spherical robot structures were used. Lead-through teaching modes became common, and sometimes sophisticated sensors are employed to maintain process consistency.[7] Experience has shown that when properly selected and implemented, these robotic applications usually lead to reduced overall manufacturing costs and improved product quality compared with manual method.

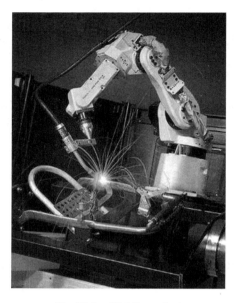

Fig.17.1　Welding robot

Assembly operations. The most advanced level of technology employing third-generation industrial robots is found in assembly. System repeatability is of utmost importance. End-of-arm tooling must be compliant, i.e., have both force and displacement control to adjust part insertions, which require that the robot actually "feel" its way along. [8] This technology usually requires a measure of artificial intelligence. Assembly robots generally are electronically driven and operate in clean environments.

Other applications. Other typical applications of robots include inspection, quality control, and repair; processing laser and water jet cutting and drilling, riveting, and clean room operations; and applications in the wood, paper, and food-processing industries. As industrial robot technology and robot intelligence improve even further, additional applications may be justified effectively.

17.5　Medical Robots

In recent years, the field of medicine has been also invaded by robots, not to replace qualified personnel such as doctors and nurses, but to assist them in routine work and precision tasks. Medical robotics is a promising field that really developed in the 1990s. Since then, a wide variety of medical applications have emerged: laboratory robots, telesurgery, surgical training, remote surgery, telemedicine and teleconsultation, rehabilitation, help for the deaf and the blind, and hospital robots. Medical robots assist in operations on heart-attack victims and make possible the millimeter-fine adjustment of prostheses. There are, however, many challenges in the widespread implementation of robotics in the medical field, mainly due to issues such as safety, precision, cost and reluctance to accept this technology.

17.6 Underwater Robots

More than 70% of the earth is covered by ocean. However, little effort has been made to utilize or protect this vast resource, compared with space or terrestrial programs. During the last few years, the use of underwater robotic vehicles has rapidly increased, since such vehicles can be operated in the deeper, riskier areas that divers cannot reach. The potential applications of such vehicles include fishing, underwater pollution monitoring, rescue, and waste cleaning and handling in the ocean as well as at nuclear sites. Most commercial unmanned underwater robots are tethered and remotely operated, referred to as remotely operated vehicles (ROVs). However, extensive use of manned submersibles and ROVs is currently limited to a few applications because of very high operational costs, operator fatigue and safety issues.

17.7 Walking Robots

There has been great effort in studying mobile robots that use legs as their locomotion system. Some developments are shown in Fig.17.2. The legs of walking robots are based on two or three degrees-of-freedom (DOF) manipulators.

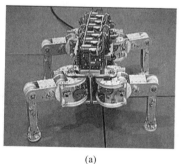
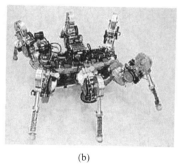

(a)   (b)   (c)

Fig.17.2   (a) Titan-VIII (Photograph courtesy of Tokyo Institute of Technology);
          (b) Lauron III (Photograph courtesy of FZI Forschungszentrum Informatik);
          (c) SILO4 (Industrial Automation Institute—CSIC)

Movement on legs confers walking robots certain advantages as opposed to other mobile robots:

1. Legged robots can negotiate irregular terrain while maintaining their body always leveled without jeopardizing their stability;
2. Legged robots boast mobility on stairs, over obstacles and over ditches as one of their main advantages;
3. Legged robots can walk over loose and sandy terrain;
4. Legged robots have inherent omnidirectionality;
5. Legged robots inflict much less environmental damage than robots that move on wheels or tracks.

17.8 Humanoid Robots

When talking about dynamically stable walking robots, humanoid robots come to mind. Actual autonomous biped robots did not appear until 1967, when Vukobratovic *et al.* led the first experiments with dermato-skeletons. The first controller-based biped robot was developed at Waseda University, Tokyo, Japan, in 1972. The robot was called WL-5.

Although the first bipeds robots were highly simplified machines under statically stable control, later developments have yielded truly sophisticated, extremely light, skillful robots (see Fig.17.3).[9] These novel developments have fed a huge amount of research that can be grouped into three major research areas: gait generation, stability control, and robot design.

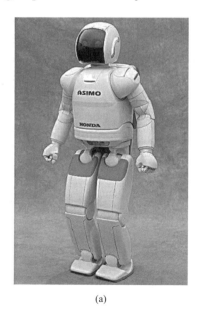  

(a) (b) (c)

Fig.17.3　Latest biped robots. (a)Photograph of ASIMO courtesy of American Honda Motor Co.;
(b) Photograph of HRP-2 courtesy of Kawada Industries, Inc.;
(c) Photograph of QRIO courtesy of Sony Entertainment Robot Europe

Research in humanoid robotics is currently shifting from locomotion issues to interaction between humans and robots. The dexterity of ASIMO, QRIO, and HRP-2 for moving up and down stairs, sitting down and standing up and dancing makes it difficult for biped-locomotion researchers to keep at the summit of legged-robotics research.[10] New trends in humanoid robotics research consider the robot's ability to interact with humans safely and the robot's ability to express emotions. The final goal will be to insert humanoid robots into the human environment, to assist the elderly and the disabled, to entertain children and to communicate in a natural language.

17.9 Intelligent Robots

Intelligent robot is a mechanical system. Compared with traditional robots, it has comprehensive improvements in perception, decision-making and performance and can simulate human behavior, emotions and thinking.[11] With a smart "brain", it can follow the instructions of humans, accomplish tasks according to the preset programs, learning and improving their behaviors while interacting with human beings. Such highly intelligent robots are widely applied to different fields, arousing people's infinite interest and rich imaginations, because they can release people from dangerous, harmful, repeated and strenuous work environments, as well as those that are beyond human capabilities in the sky or deep sea, or demanding high precision and specialized work conditions.

Considering from the application environment, intelligent robot can be divided into the industrial robots, service robot and specialized robots. Industrial robot is a self-controlled and self-powered machinery device that can move automatically and finish various tasks. After receiving instructions from humans, this industrial robot will design the path of movement and then begin to operate according to predetermined procedures, including welding, coating, assembly, picking and placing objects (such as packaging, palletizing), and can also detect and test products, which have been described in detail in the earlier part of this chapter, and will not be repeated here.

A service robot is a robot that "performs useful tasks for humans or devices, but does not include industrial automation applications". For example, domestic servant robots, such as intelligent sweeping robots and window cleaning robots, can act as human assistants to do housework. Other types of service robots are home socialized robots, companion robots, mobility assistance robots, and pet-sitting robots. They can communicate with humans and perform assigned tasks such as caring for the elderly and children, reminder services and residential patrols.

In addition, some robots, such as emotional communication robots, children's education robots, intelligent robot learning platforms, private unmanned aerial vehicles (UAVs), personal mobile robots, humanoid robots, robotic pets, etc., can communicate information and emotions with humans through interactive voice control technology. [12]

Commercial service robots include reception robots, shopping guide robots, cooking robots, office robots, and security robots. Personalized robot products can also be customized according to the application environments, which can complete tasks such as advertising, providing advice and guidance, assisting office work, and performing security patrols, etc.

Boston Dynamics is devoted to developing advanced robot with high maneuverability, flexibility and fast moving speed. By making full use of sensors and dynamic control, it developed the BigDog for all-terrain goods transportation, the double-feet robot Atlas with ultra-high balancing ability and the handle with wheel-leg and ultra-strong leaping ability (see Fig.17.4).

Fig.17.4 Representative works of Boston Dynamics

Intelligent robot industry sets a significant standard to evaluate the technological innovation and high-end manufacturing level of a country. Its development is attracting more and more attentions in the world. In order to seize development opportunities and gain competitive edges in high-tech fields represented by robot, major economic entities in the world have successively formulated national strategies on robot industry development, and realized a series of breakthroughs in robotics technologies.

With the development of intelligent hardware and artificial intelligence technology, intelligent robots have achieved outstanding progresses and are widely used in many fields. Diversified user demands and increasing emergence of new technologies drive robot to develop towards a highly-intelligent, highly-adaptive and network-based prospect.[13]

Notes

[1] A robot is a reprogrammable multifunctional manipulator designed to move material, parts, tools, or specialized devices through variable programmed motions for the performance of a variety of tasks.

句意：机器人是一种可编程的多功能操纵装置，通过可编程运动来完成多种作业，如运送材料、零件、工具或其他专用设备。

[2] Consequently, we will describe the components of a robot system from the point of view of information transfer.

句意：因此，我们将根据信息传递的观点来描述机器人系统的组成部分。

[3] We could easily have labeled the computation module "computer", as most of the functions such as servo, kinematics, dynamics and workplace sensor analysis are typically performed by digital computers.

句意：我们可以很容易地将计算模块称为"计算机"，因为大部分功能诸如伺服控制、运动学、动力学和工作现场的传感器分析等，通常都是由数字计算机来完成的。

[4] Variable-sequence Robot. Robot that can be programmed for a given sequence of operations and can easily be changed or reprogrammed.

句意：可变顺序机器人。可以对这类机器人进行编程，使其按一定的顺序工作，可以很容易地改变这种顺序或者重新编程。

[5] Intelligent Robot. Robot that uses sensory perception to evaluate its environment and make decisions and proceeds to operate accordingly.

句意：智能机器人。这种机器人采用感官知觉对它周围的环境进行评价和做出决定，并据此进行工作。

[6] Replacing unskilled human labor often in hazardous jobs, these robots had to be robust and low in initial and maintenance costs.

句意：这些机器人通常被用来替代从事危险性工作的非技术工人，它们必须坚固耐用而且具有较低的使用费用和维护费用。

[7] Lead-through teaching modes became common, and sometimes sophisticated sensors are employed to maintain process consistency.

句意：示教型这种方式变得比较常见，有时还需要采用复杂的传感器来保持过程的一致性。

[8] End-of-arm tooling must be compliant, i.e., have both force and displacement control to adjust part insertions, which require that the robot actually "feel" its way along.

句意：臂端工具必须具有柔性功能，也就是说，它们具有受力与位移控制功能，从而调节操作部件的介入，这就要求机器人能实际"感觉"它的运动路径。

[9] Although the first biped robots were highly simplified machines under statically stable control, later developments have yielded truly sophisticated, extremely light, skillful robots.

句意：尽管在静态稳定控制之下，第一个双足机器人的机构得到了高度简化，但后期的发展产生了真正精密、极度轻巧、技术熟练的机器人。

[10] The dexterity of ASIMO, QRIO, and HRP-2 for moving up and down stairs, sitting down and standing up and dancing makes it difficult for biped-locomotion researchers to keep at the summit of legged-robotics research.

句意：ASIMO、QRIO 和 HRP-2 能够灵巧地上下楼，坐下与起立，还能跳舞，这对双足行走机器人的研究人员试图保持他们在有腿机器人研究方面的尖端水平提出了严峻挑战。

[11] Intelligent robot is a mechanical system. Compared with traditional robots, it has comprehensive improvements in perception, decision-making and performance and can simulate human behavior, emotions and thinking.

句意：智能机器人是一个机械系统，它与传统机器人相比，在感知、决策与性能上做了全面改进，从而可以模拟人类的行为、情感和思维。

[12] In addition, some robots, such as emotional communication robots, children's education robots, intelligent robot learning platforms, private unmanned aerial vehicles (UAVs), personal mobile robots, humanoid robots, robotic pets, etc., can communicate information and emotions with humans through interactive voice control technology.

句意：此外，有些机器人，如情感交流机器人、儿童教育机器人、智能机器人学习平台、私人无人驾驶飞行器、个人移动式机器人、人形机器人、机器人宠物等，都可以通过交互式音控技术与人类进行信息和情感交流。

[13] With the development of intelligent hardware and artificial intelligence technology, intelligent robots have achieved outstanding progresses and are widely used in many fields. Diversified user demands and increasing emergence of new technologies drive robot to develop towards a highly-intelligent, highly-adaptive and network-based prospect.

句意：随着智能软件和人工智能技术的不断发展，智能机器人已经获得了显著进展，被广泛用于多个领域。各式各样的用户需求和不断涌现的新技术正推动机器人朝着高度智能化、高度自适应性和网络化的前景发展。

Glossary

actuator   *n.* 驱动器，执行机构
all-terrain   *adj.* 全地形的
anthropomorphically   *adv.* 类人地，从人体结构上
articulated robot   关节型机器人
autonomous   *adj.* 自治的，自主的
boast   *v.* 以……而自豪，自夸
complement   *n.* 补充，互补
compliant   *adj.* 顺从的，依从的，柔性的
dedicated   *adj.* 专用的
dexterity   *n.* 灵巧，机敏
domestic servant robot   家庭服务机器人
electromyogram   *n.* [物]肌动电流图
electrooculogram   *n.* [医]眼动电图，眼电（流）图
end effector   末端执行器
end-of-arm tooling   臂端工具
envisage   *v.* 正视，想象，展望
fixed-sequence robot   固定顺序机器人
gripper   *n.* 夹持器，抓爪
hard-automation   刚性自动化
hierarchical   *adj.* 分层的，分级的
home socialized robot   家庭社会化机器人
humanoid   *adj.* 类人的，有人类特点的
implementation   *n.* 执行，履行，落实，供给工具
invate   *v.* 侵入，侵占
jeopardize   *v.* 危及
knowledge-based   基于知识的
lead-though teaching   示教的，仿效的
locomotion   *n.* 运动，移动
manipulator   *n.* 机械手，操作器
multifunctional   *adj.* 多功能的
orientation   *n.* 定位，方位
palletize   *v.* 码垛，堆积
partition   *v.* 分割
peripheral equipment   周边设备，外围设备

pet-sitting robot　　照看宠物机器人
playback robot　　再现式机器人
prearrange　　*v.* 预先安排，预定
precision　　*n.* 精确（性），精度
private unmanned aerial vehicle　　私人无人驾驶飞行器
prosthesis　　*n.* 假肢
proximity sensor　　接近传感器
reprogrammable　　*adj.* 可重复编程的
robust　　*adj.* 健壮的，坚固的
security robot　　安保机器人
sensory perception　　感官的感觉，感官知觉
social-demand　　社会需求
sophisticate　　*v.* 使复杂，使精致
sophisticated　　*adj.* 尖端的，高级的；复杂的；久经世故的
strenuous　　*adj.* 费力的，麻烦的
submersible　　*adj.* 潜水的，水下的
surgical　　*adj.* 外科医生的，外科用的，外科手术的
teach box　　示教盒
tether　　*v.* 用绳索牵系
teleconsultation　　*n.* （用遥测设备、闭路电视等进行的）远距离会诊
telemedicine　　*n.* （通过遥测、电话、电视等手段求诊的）远距离医学
terrestrial　　*adj.* 陆地的
terrain　　*n.* 地域，地势，地形
trajectory　　*n.* 轨迹，路线
tutelage　　*n.* 监护，指导
variable-sequence robot　　可变顺序机器人
zero reference point　　零参考点，基准零点

Unit 18 Additive Manufacturing

18.1 Introduction

What is additive manufacturing? Additive manufacturing (AM), formerly known as Rapid Prototyping, is one of the rapidly developed advanced manufacturing technologies over the last few decades. The ASTM International Committee F42 on Additive Manufacturing technology defines *additive manufacturing* as the process of joining materials to make objects from 3D model data, usually layer upon layer, as opposed to subtractive manufacturing (e.g., turning, milling, grinding, EDM, etc.) methods.[1]

3D printing is defined as the fabrication of objects through the deposition of a material using a print head, nozzle, or other printer technology. However, this term is often used synonymously with additive manufacturing since just as its name implies. 3D printing means successive layers of materials are spread and then built up into a three-dimensional solid object. So, 3D printing is now widely used to represent the newly developed manufacturing technology *additive manufacturing*.[2]

Additive manufacturing is the official industry standard term according to ASTM and ISO, but 3D printing has become the *de facto* standard term, and has become more popular than AM.

AM is used to build physical models, prototypes, patterns, tooling components, and production parts in plastic, ceramic, glass, and composite materials. AM systems use thin cross section from 3D models created by CAD software, 3D scanning systems, medical scanners, and video games.

Additive manufacturing encompasses seven distinctly different processes. Parts can be created in a layer-by-layer fashion by extruding, jetting, photo-curing, laminating, or fusing materials. Materials include plastics, metals, ceramics, and composites.

AM offers a distinctive form of manufacturing that is "toolless" and capable of producing extremely complex shapes and geometric features. The direct production of parts using AM facilitates small batch sizes, custom parts, assembled components, lightweight, and complex internal structures. AM is far from replacing conventional processes for most products, yet for some product categories, AM production has become competitive. For example, AM is used to manufacture about 90% of the world's plastic shells for custom in-the-ear hearing aids. More than an estimated 10 million have been produced in this way.

18.2 AM Processes and Materials

AM processes. Additive manufacturing encompasses many processes and materials. The range of possibilities can be confusing for newcomers to the AM industry. Also, AM system manufacturers have created unique process names and material designations to differentiate themselves from their competitors. Many of the "different" systems actually employ similar processes and share similar, or identical materials. Fortunately, the industry has a standardized system for grouping AM processes and materials and categorizing them into families.

AM processes have a lot in common. Specially, input to the system consists of 3D model data, and fabrication occurs by joining materials in successive layers. However, a wide range of quite different processes is available under the AM umbrella. Therefore, defining different AM processes so that the available systems fall into specific process categories is useful and instructive.

In January 2012, the ASTM International Committee F42 on Additive Manufacturing Technologies approved the list of AM process category names and definitions in the specification "Standard Terminology for Additive Manufacturing Technologies".

- Material extrusion—an additive manufacturing process in which material is selectively dispensed through a nozzle of orifice, e.g., Fused deposition modeling or FDM (Fig.18.1).

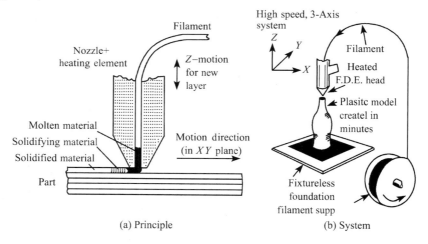

(a) Principle　　　　　　　　　(b) System

Fig.18.1　Fused deposition modeling

- Material jetting—an additive manufacturing process in which droplets of build material are selectively deposited.
- Binder jetting—an additive manufacturing process in which a liquid bonding agent is selectively deposited to join powder materials, e.g., 3D printing (Fig.18.2).

Note that this 3D printing shown in Fig.18.2 only represents in a traditional sense the AM process that a layer of powder material is spread on the previously paved powder, and then an inkjet printing head projects droplets of binder material onto the powder to build AM parts, which is not the same as the inclusive term 3D printing used today.

- Sheet lamination—an additive manufacturing process in which sheets of material are bonded to form an object, e.g., Laminated object manufacturing or LOM (Fig.18.3).

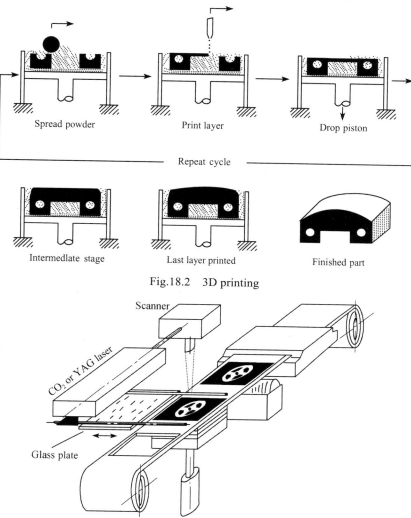

Fig.18.2　3D printing

Fig.18.3　Laminated object manufacturing

- Vat photopolymerization—an additive manufacturing process in which liquid photopolymer in a vat is selectively cured by light-activated polymerization, e.g., Stereolithography or SL (Fig.18.4).
- Powder bed fusion—an additive manufacturing process in which thermal energy selectively fuses regions of a powder bed, e.g., Selective Laser Sintering or SLS (Fig.18.5 and Fig.18.6).

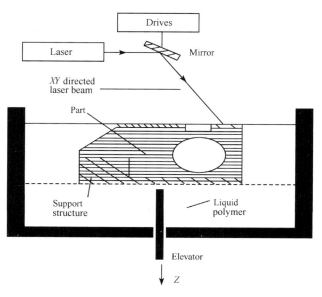

Fig.18.4　Stereolithography

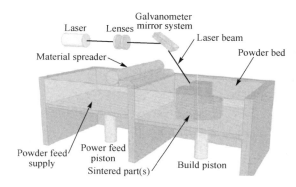

Fig.18.5　Selective Laser Sintering

Fig.18.6　Cabin bracket made with SLS

- Directed energy deposition—an additive manufacturing process in which focused thermal energy is used to fuse materials by melting as the material is being deposited.

Materials. The two major categories of AM materials are plastics and metals. A variety of filled and composite materials are also available, as well as ceramics and ceramic-metal hybrids.

Plastics are classified into two groups based on their behavior at high temperatures. Thermoplastics retain their properties and can be repeatedly melted, cooled and hardened, and melted again. Thermoset plastics are permanently "set" once they are formed and cannot be remelted.[3]

The plastic materials used in material extrusion systems are exclusively thermoplastics, such as ABS, polycarbonate (PC), polyethylene terephthalate (PET), polyvinyl alcohol (PVA), nylon, and poly lactic acid (PLA). The materials used in photopolymer processes are thermoset plastics. These are typically proprietary acrylics, or epoxies. Most of these liquid resins are formulated to photocure when exposed to ultraviolet energy, some also photocure when exposed to energy in the visible light spectrum.

With the development of AM technology, some metal materials have also been used to build functional parts. The growing number of metal materials available for additive manufacturing systems is impressive. A designer can choose from a wide range, including:

- Tool steels
- Stainless steels
- Commercially pure titanium
- Titanium alloys
- Aluminum alloys
- Nickel-based alloys
- Cobalt-chromium alloys
- Copper-based alloys
- Gold
- Silver

18.3 Applications of Additive Manufacturing

The first commercial systems were on the market in the late 1980s. Now industrial/business machines constitute the leading industrial sector, although it dropped 1.0% from 2013 to 2014 (Fig.18.7). The aerospace sector grew 2.5% over 2013, and the academic institutions and government/military sectors also grew. The consume products/electronics category covers a broad range of products, including mobile phones, home electronics and computers, kitchen appliances, tools, and toys. These industries typically produce parts in large volumes, and the product life circles are relatively short. AM accelerates product development by enabling rapid design iteration and optimization for companies in these industries.

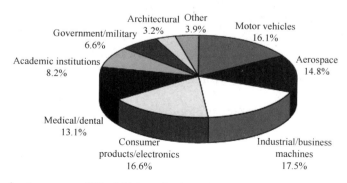

Fig.18.7 Approximate revenues (%) of AM service in various industrial sectors (Wohlers Report 2014)

Large manufacturers in the automotive industry invested in AM machines shortly after they were commercialized, and adopted AM for prototyping and rapid product development.

The motor vehicle industry continues to use AM in product development, although production volumes are typically too high to use AM for most final part applications.

The medical industry has adopted AM for models, surgical cutting and drill guides, and orthopedic implants. An estimated 20 different AM medical implant products have gained permit from the Food and drug administration (FDA). They range from cranial implants to hip, knee, and spinal implants. More than 100000 acetabular (hip cup) implants have been produced, and about 50000 of them have been implanted into patients.

The aerospace industry was an early adopter of AM and began to explore applications soon after the technology was commercialized. Boeing and Bell Helicopter began to use polymer AM parts for non-structural production applications in the mid 1990s. Boeing has produced tens of thousands of AM parts, representing more than 200 unique parts on 15 different commercial and military aircraft.

Modeling and prototyping were the first applications for parts made on AM systems (Fig.18.8 and Fig.18.9). Visual aids, presentation models and fit and assembly models account for more than third (36.8%) of all applications. Also, a portion of the "functional parts" category is comprised of functional prototypes—parts that are used for rigorous testing. Models and prototypes are used to communicate design and clarify ambiguities from engineering drawings. Also, they are used as presentation models in meetings with management and potential clients and as planning models for complex procedures such as surgery. Prototyping is an important step in new product development, allowing designers to quickly iterate and optimize their designs. [4]

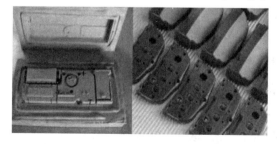 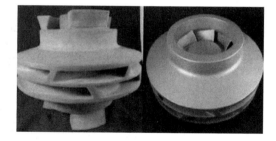

Fig.18.8 Silicone rubber mold and urethane part   Fig.18.9 Investment casting pattern & cast part

The making of master patterns for prototype tooling is another important application of AM systems. Silicon rubber is poured over an AM master to create a rubber tool, which is then used to cast urethane parts. AM parts are also used as patterns for investment casting of metal parts. This eliminates the need for expensive and time-consuming wax pattern tooling.

A number of AM materials can be burned out of a ceramic shell for the investment casting process. AM processes can also be used to produce many other types of tools, including jigs, fixtures, templates, gauges, and drill and cutting guides. They can be expensive and time-consuming to produce using conventional manufacturing methods. After a successful test of a material extrusion 3D printer on the International Space Station, NASA and the European Space Agency are investigating metal AM in space (Fig.18.10). Several food 3D printers are in development and reaching the market. Bioprinting is also a rapidly developing field (Fig.18.11). Current research in the 3D printing of human body living issues suggests a future where body parts and possibly even entire organs will be manufactured from a patient's own cell (stem cell). [5]

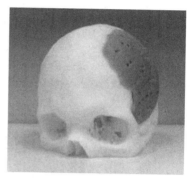

Fig.18.10　GE aviation metal parts　　　Fig.18.11　Cranial implant made in OXPEKK

AM technology has been introduced successfully in the industries of automotive, aerospace, shipbuilding, IT equipment, electrical appliances and jewelries (Fig.18.12). Microelectronic-mechanical system will become the other important field of application with AM. The height of microelectronic circuits and microelectronic-mechanism is typically between 2 and 10 mm. That means AM processes will be a feasible way of fabricating these systems.

18.4　Conclusions

Product features, quality, cost and time to market are important factors for a manufacturer to remain competitive. AM systems offer the opportunities to make products faster, and usually at lower costs than using conventional methods. Since AM can substantially reduce the product development cycle time, more and more businesses are taking advantage of the speed at which product design generated by computers can be converted into accurate models that can be held, viewed, studied, tested and compared.

This is a rapid development area. Capacities and the potential of AM technologies have attracted a wide range of industries to invest in these technologies. It is expected that greater effort is needed for research and development of those technologies so that they will be widely used in product-oriented manufacturing industries.

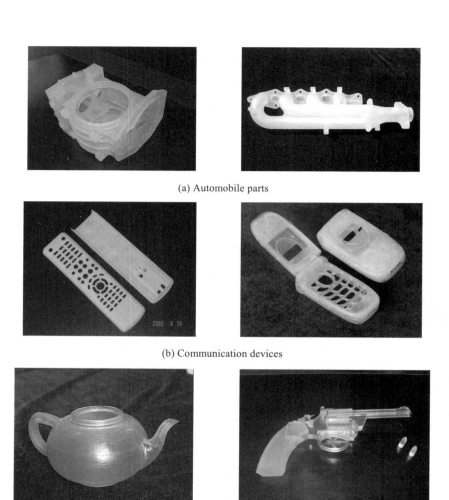

(a) Automobile parts

(b) Communication devices

(c) Artifacts

(d) Jewelry patterns

Fig.18.12 SL prototypes

Notes

[1] The ASTM International Committee F42 on Additive Manufacturing technology defines additive manufacturing as the process of joining materials to make objects from 3D

model data, usually layer upon layer, as opposed to subtractive manufacturing (e.g. turning, milling, grinding, EDM, etc.) methods.

句意：美国材料与试验学会 F42 增材制造技术国际委员会将增材制造定义为一种将材料根据三维模型数据采用逐层连接方式制作物件的工艺过程，它完全不同于传统的材料去除制造方法，例如，车削、铣削、磨削和电火花加工等。

[2] However, the term is often used synonymously with additive manufacturing since just as its name implies, 3D printing means successive layers of materials are spread and then built up into a three-dimensional solid object. So, 3D printing is now widely used to represent the newly developed manufacturing technology additive manufacturing.

句意：然而，这个术语常常被用作增材制造的同义词。因为顾名思义，3D 打印就是将某些材料逐层铺放然后堆积制成三维实体物件。所以 3D 打印现在就被广泛用来代表新开发的制造技术——增材制造。

[3] Thermoplastics retain their properties and can be repeatedly melted, cooled and hardened, and melted again. Thermoset plastics are permanently "set" once they are formed and cannot be remelted.

句意：热塑性塑料加热后可反复熔融、冷却、固化和重熔而保持它的性能。而热固性塑料一旦成形就永久不变，无法再重熔。

[4] Prototyping is an important step in new product development, allowing designers to quickly iterate and optimize their designs.

句意：原型制作在新产品开发过程中是非常重要的一步，它使设计师能快速反复修改和优化他们的设计。

[5] Current research in the 3D printing of human body living issues suggests a future where body parts and possibly even entire organs will be manufactured from a patient's own cell (stem cell).

句意：当前关于人体活性组织的 3D 打印技术为我们展示了这样的前景：将来人体部件，甚至可能整个器官都可以从病人自身的细胞（干细胞）中制造出来。

Glossary

acrylonirile butadine styrene (ABS)　丙烯腈丁二烯苯乙烯共聚物（ABS 塑料）
additive manufacturing　增材制造
acetabular implant　髋臼植入体
ambiguity　*n.* 歧义
binder　*n.* 黏合剂，黏接剂

biocompatible　*adj.* 生物相容的，不会引起排斥的

biodegradable　*adj.* 可生物降解的

bioprinting　*n.* 生物打印

ceramic-metal hybrid　陶瓷金属混合物

commercialize　*vt.* 使商业化；使商品化

curing　*n.* （化学）固化，硬化

de facto　<拉>事实上，事实上的，既成事实的，未经合法手续批准的

directed energy deposition　直接能量沉积

droplet　*n.* 液滴

enabling　（enable 的现在分词）使能够，赋能给

extrusion　*n.* 挤出，推出

fixture　*n.* 夹具

fused deposition modeling (FDM)　熔融沉积造型

fusion　*n.* 融合；熔化 熔接

gauge　*n.* 计量器，量规，仪表

hybrid　*n.* 杂种，混合，杂交　*adj.* 混合的，杂交的

in-the-ear hearing aids　耳内助听器

investment casting　熔模铸造

iteration　*n.* [数] 迭代，反复，重复

jetting　*n.* 喷射，注射

jewelry　*n.* 珠宝，首饰

jig　*n.* 夹具，钻模

laminated object manufacturing (LOM)　迭层制造

living issue　活体组织

marketing　*n.* 行销，销售，市场营销

microvessel　*n.* 微脉管，微血管

master pattern　原模型，母模

nozzle　*n.* 喷嘴，管口，排气口

orthopedic implant　整形外科植入体

photocure　*vt.* 光固化

powder bed fusion　粉床融合

polycarbonate　*n.* 聚碳酸酯

polymerize　*vt.* 使……聚合　*vi.* 聚合

polymide (nylon)　*n.* 尼龙，聚酰胺

polyethylene terephthalate (PET)　*n.* 对苯二酸盐；对苯二酸酯

polyvinyl alcohol (PVA)　聚乙烯醇

poly lactic acid (PLA)　聚乳酸
photopolymerization　*n.* 光聚合，光致聚合作用
prototyping　*n.* 原型（样件）制作，样机制作
rapid prototyping　快速原型制造
selective laser sintering (SLS)　选区激光烧结
sheet lamination　薄片叠层（制造）
shipbuilding　*n.* 造船，造船业
silicon rubber　硅橡胶
spectrum　*n.* 光谱，频谱，范围
spinal implant　脊椎植入体
synonymously　*adv.* 同义地，同等含义地（表达）
Stereolithography (SL)　*n.* 光固化立体造型
subtractive manufacturing　减材制造
template　*n.* 模板，样板
thermoplastic　*adj.* 热塑性的　*n.* 热塑性塑料
thermoset　*adj.* 热固性的
three dimensional printing (3D-printing)　三维打印
tooling　*n.* 工艺装备，工模具
ultraviolet　*adj.* 紫外的，紫外线的　*n.* 紫外线辐射，紫外光
video games　电子游戏
rigorous　*adj.* 严格的，严厉的，严酷的
visual aids　视觉教具

Unit 19　New Energy

19.1　Introduction

Energy constitutes the four pillars of a civilized society together with new materials, biotechnology and information technology. Energy is the main material basis to promote social development and economic progress, and every progress in energy technology has driven the development of human society. With the consumption of non-renewable fossil fuel resources such as coal, oil and natural gas and the need for ecological environmental protection, the development of new energy will promote the transformation of the world's energy structure, and the maturity of new energy technologies will bring revolutionary changes in the industrial field.

19.2　Classification of Energy Resources

Energy has a variety of classification methods, according to the form, it can be divided into primary energy (such as coal, oil, natural gas, solar energy, etc.) and secondary energy (electricity, gas, steam, etc.); according to the circulatory modes, it can be divided into non-renewable energy (fossil fuel) and renewable energy (biomass energy, hydrogen energy, chemical energy); according to the nature of energy generation, it can be divided into energy-containing resource (coal, oil, etc.) and process energy (solar energy, electric energy, etc.); [1] according to the requirements of environmental protection, energy can be divided into clean energy (also known as green energy, such as solar energy, hydrogen energy, wind energy, tidal energy, etc., including waste disposal, etc.) and non-clean energy; if according to the maturity of the present stage, it can be divided into conventional energy and new energy.

19.3　New Energy

New energy (NE): also known as unconventional energy. It refers to various forms of energy other than traditional energy resources. It refers to the energy that has just begun to be developed and utilized or is under active research and needs to be promoted, such as solar energy, geothermal energy, wind energy, ocean energy, biomass energy and nuclear energy.

In 1981, the United Nations Conference on New and Renewable Energy defined new energy technology as: on the basis of new technologies and new materials, make traditional renewable energy resources obtain modernized development and utilization, to replace fossil energy that have limited resources and pollute the environment with inexhaustible renewable energy resources, focusing on the development of solar energy, wind energy, biomass energy, tidal energy, geothermal energy, hydrogen energy and nuclear energy (atomic energy).[2]

New energy generally refers to the renewable energy developed and utilized on the basis of new technologies, including solar energy, biomass energy, wind energy, geothermal energy, wave energy, ocean current energy and tidal energy, as well as the thermal cycle between the ocean surface and deep layer. In addition, there are hydrogen energy, biogas, alcohol, methanol, etc., and coal, oil, natural gas, water energy and other widely used energy resources are called conventional energy resources. With the limitation of conventional energy and the increasingly prominent environmental problems, new energy with environmental protection and renewable characteristics has been paid more and more attention by all countries.

19.4 Solar Energy

Solar energy generally refers to the sun radiation energy. The main use of solar energy has three forms: solar energy photothermal conversion, photoelectric conversion and photochemical conversion. The main methods of using solar energy are solar cells which convert the energy contained in sunlight into electricity through photoelectric conversion and solar water heaters which use the heat of sunlight to heat water, and use hot water to generate electricity.[3]

Photovoltaic power generation is a technology using the photovoltaic effect on the semiconductor interface to directly convert light energy into electrical energy. It is mainly composed of solar panels (components), controllers and inverters. After the solar cells are packaged and protected in series, solar cell modules with a large area can be formed, and then combined with the power controller and other components to form a photovoltaic power generation device. Another solar energy technique is solar photothermal technique, modern solar photothermal technology focuses sunlight and makes it produce hot water, steam and electricity.

19.5 Nuclear Energy

There are three main forms of nuclear energy release.

A. Nuclear fission energy

The so-called nuclear fission energy is the energy released by the fission of some heavy nuclei (such as uranium-235, plutonium-239, etc.).

B. Nuclear fusion energy

By two or more hydrogen nuclei (such as hydrogen isotopes—deuterium and tritium) combined into a heavier nucleus, release a huge energy with mass loss at the same time, this reaction is called nuclear fusion reaction and the energy released is called fusion energy.[4]

C. Nuclear decay

Nuclear decay is a natural, much slower form of fission that is difficult to exploit because of its slow release of energy.

Since the birth of the first nuclear power plant in the 1950s, the global nuclear fission power generation has developed rapidly, nuclear power technology has been continuously improved, and various types of reactors have emerged, such as pressurized water reactor, boiling water reactor, heavy water reactor, graphite reactor, gas cooled reactor and fast neutron reactor. It is very difficult for humans to achieve nuclear fusion and control it, and plasma is the most promising way to achieve nuclear fusion reaction. By heating the plasma to the ignition temperature, certain devices and methods are used to control the density of the reactants and the time to maintain this density, and the controlled nuclear fusion can be realized.

19.6　Ocean Energy

Ocean energy refers to all kinds of renewable energy contained in marine water, including tidal energy, wave energy, ocean current energy, seawater temperature difference energy, seawater salinity difference energy, etc. These energy resources are renewable and do not pollute the environment. So ocean energy is a new energy urgently needed to be developed and utilized with strategic significance.

19.7　Wind Energy

Wind energy is formed by its flow due to solar radiation. Compared with other energy resources, wind energy has obvious advantages: its reserves are large, 10 times that of water energy, widely distributed, never exhausted, and it is particularly important for islands and remote areas with inconvenient transportation and far from the main grid. Wind power generation is the most common form of contemporary use of wind energy. Since the end of the 19th century when Denmark invented a wind turbine, people realized that oil and other energy will be exhausted and began to pay attention to the development of wind energy. In 1977, the Federal Republic of Germany built one of the world's largest power-generating windmills. The windmill is 140 meters high, each blade is 40 meters long, weighs 18 tons and is made of fiberglass.

Wind power generation technology involves aerodynamics, automatic control, mechanical transmission, electrical machinery, mechanics and materials science and other multidisciplinary comprehensive high-tech system engineering.[5] At present, in the field of wind power generation, the research difficulties and hot spots mainly focus on the large-scale wind turbine, the advanced control strategy and optimization technology of wind turbine.

19.8 Biomass Energy (Aka Bioenergy)

Biomass energy comes from biomass, which is also a form of energy stored in living organisms in the form of chemical energy, and directly or indirectly comes from the photosynthesis of plants. Biomass energy belongs to a stored solar energy and is the only renewable source of carbon that can be converted into conventional solid, liquid or gaseous fuels. Biomass energy uses organic matter (such as plants) as fuel, through gas collection, gasification, combustion and digestion (wet waste only) and other technologies to produce energy.

The development and utilization of biomass energy has been highly valued in many countries, biomass energy is likely to become the main member of the future sustainable energy system. The development of biomass energy technology includes biomass gasification technology, biomass curing technology, biomass pyrolysis technology, biomass liquefaction technology, biodiesel technology and biogas technology, etc.[6]

19.9 Geothermal Energy

Geothermal heat source in the interior of the earth comes from molten magma, and energy release of the decay of some radioactive elements, etc. Spontaneous thermonuclear reactions of radioactive materials are the main heat source of the earth. China is rich in geothermal resources and widely distributed, with 5500 geothermal hot spots and 45 geothermal fields, with a total geothermal resources of about 3.2 million megawatts.

19.10 Hydrogen Energy

The advantages of hydrogen energy: hydrogen energy is a safe and environmentally friendly energy. Hydrogen molecular weight is 2, only 1/14 of the air, therefore, hydrogen leakage in the air will automatically escape from the ground, will not form aggregation. Other fuel and gas will gather on the ground and pose a flammable and explosive risk. Hydrogen is tasteless and non-toxic, will not cause human poisoning, its combustion product is only water, does not pollute the environment. The calorific value of 1 kg of hydrogen is 34000 kcal, which is three times that of gasoline. Oxyhydrogen flame temperature up to 2800 degrees, higher than conventional liquid gas. The flame is straight, with small heat loss and high utilization efficiency.

Hydrogen energy utilization technology includes hydrogen production technology, hydrogen purification technology and hydrogen storage and transport technology. The range of hydrogen production technology is very wide, including fossil fuel hydrogen production, electrolytic water hydrogen production, solid polymer electrolyte electrolytic hydrogen production, high temperature water vapor electrolysis hydrogen production, biological hydrogen production, biomass hydrogen production, thermochemical decomposition of water hydrogen production, methanol reforming hydrogen production and H_2S decomposition hydrogen production.[7] Hydrogen storage is an important guarantee for hydrogen energy utilization, and the main technologies include liquefied hydrogen storage, compressed hydrogen storage, metal hydride hydrogen storage, coordination hydride hydrogen storage, physical adsorption hydrogen storage, organic hydrogen storage and glass microsphere hydrogen storage. The main applied technologies of hydrogen include: fuel cell, gas turbine (steam turbine) power generation, MH/Ni batteries, internal combustion engines and rocket engines.

19.11　Development Direction of New Energy Vehicles

The car as a means of transportation discharges a large amount of carbon, nitrogen, and a variety of air pollutants every day, such as sulfur oxides, hydrocarbons and lead compounds, resulting in atmospheric pollution, and serious harm to human health and ecological environment.[8] Energy saving and emission reduction is the eternal theme of the development of the automobile industry, and constantly strengthening the work of energy saving and emission reduction has become an urgent need for China's economy to achieve sound and rapid development. In developed countries, car determines the demand for oil, is also the key factor that affects greenhouse gas and harmful gas emissions, and realize the goal of environmental protection to reduce oil consumption and emissions of car. But on the other hand, automobile industry is a pillar industry, but also the basic means of transportation, governments have to maintain the development of the automobile to promote economic development and the improvement of people's living welfare. The development of energy-saving and environmentally friendly vehicles can reduce oil consumption and protect the atmospheric environment while maintaining the growth of automobiles, so governments regard the development of energy-saving and environmentally friendly vehicles as an important part of realizing their energy and environmental policies and the sustainable development of the automobile industry.

New energy vehicles mainly refer to pure electric vehicle, hybrid electric vehicle, fuel cell electric vehicle, hydrogen engine car and hydrogen train.

A. Pure Electric Vehicle

Pure electric vehicle is a vehicle using a single battery as the energy storage power source, through the battery to provide electric energy to the motor, drive the motor to operate, so as to promote the car.

B. Hybrid Electric Vehicle

Hybrid electric vehicle refers to the vehicle whose driving system is composed of two or more single drive systems that can run at the same time, and the driving power of the vehicle is provided by a single driving system or multiple driving systems jointly according to the actual driving state of the vehicle. Generally speaking, hybrid electric vehicles refer to hybrid electric vehicles (HEV), they use traditional internal combustion engines (diesel or gasoline engines) and electric motors as power sources, and some engines have been modified to use other alternative fuels, such as compressed natural gas, propane and ethanol fuel.

C. Fuel Cell Electric Vehicle

Fuel cell electric vehicle uses hydrogen and oxygen in the air under the action of the catalyst, the electric energy generated by the electrochemical reaction in the fuel cell as the main power source to drive the vehicle. Generally speaking, the fuel cell converts chemical energy into electrical energy through the electrochemical reaction. Electrochemical reaction requires hydrogen as reductant, oxygen as oxidant, so the earliest development of fuel cell electric vehicle directly uses hydrogen as fuel, and hydrogen storage can use liquefied hydrogen, compressed hydrogen or metal hydride hydrogen storage, etc.

D. Hydrogen Engine Car

Hydrogen engine car is a car with hydrogen engine as the power source. The fuel used by general engines is diesel or gasoline, and the fuel used by hydrogen engine is the gas hydrogen. Hydrogen engine car is a real zero-emission vehicle, the emission is pure water, which has the advantages of no pollution, zero emissions and abundant reserves.

E. Hydrogen Train

Canada, the United Kingdom, the United States, Russia, South Korea, Japan and other countries have invested in the research and development and testing of hydrogen energy trains, and the world's first hydrogen train developed by Alstom for Germany began line testing in September 2016.

In terms of hydrogen locomotives, vehicle projects in the United States developed a hydrogen fuel cell shunting locomotive in 2009, weighing about 130 t, using a hybrid power system of fuel cells and lead-acid batteries, and 240 kW proton exchange membrane fuel cells as the main power output and battery charging. The battery provides auxiliary power output, and the instantaneous power rate of the whole system can exceed 1 MW. Canada Pacific (CP) Railway announced in 2020 the development of the first hydrogen fuel cell rail freight locomotive project for North America.

Notes

[1] Energy has a variety of classification methods, according to the form, it can be divided into primary energy (such as coal, oil, natural gas, solar energy, etc.) and secondary energy (electricity, gas, steam, etc.); according to the circulatory modes, it can be divided into non-renewable energy (fossil fuel) and renewable energy (biomass energy, hydrogen energy, chemical energy); according to the nature of energy generation, it can be divided into energy-containing energy source (coal, oil, etc.) and process energy (solar energy, electric energy, etc.).

句意：能源有多种分类方法，按形成方式可分为一次能源（如煤、石油、天然气、太阳能等）和二次能源（电、煤气、蒸汽等）；按循环方式可分为不可再生能源（化石燃料）和可再生能源（生物质能、氢能、化学能源）；按能量产生的本质又可分为含能体能源（煤炭、石油等）和过程能源（太阳能、电能等）。

[2] In 1981, the United Nations Conference on New and Renewable Energy defined new energy technology as: on the basis of new technologies and new materials, make traditional renewable energy resources obtain modernized development and utilization, to replace fossil energy that have limited resources and pollute the environment with inexhaustible renewable energy resources, focusing on the development of solar energy, wind energy, biomass energy, tidal energy, geothermal energy, hydrogen energy and nuclear energy (atomic energy).

句意：1981 年，联合国召开的"联合国新能源和可再生能源会议"对新能源技术的定义为：以新技术和新材料为基础，使传统的可再生能源得到现代化的开发和利用，用取之不尽、周而复始的可再生能源取代资源有限、对环境有污染的化石能源，重点开发太阳能、风能、生物质能、潮汐能、地热能、氢能和核能（原子能）。

[3] The main methods of using solar energy are solar cells which convert the energy contained in sunlight into electricity through photoelectric conversion and solar water heaters which use the heat of sunlight to heat water, and use hot water to generate electricity.

句意：利用太阳能的方法主要有太阳能电池，通过光电转换把太阳光中包含的能量转换为电能；另外就是太阳能热水器，它利用太阳光的热量加热水，并利用热水发电等。

[4] By two or more hydrogen nuclei (such as hydrogen isotopes—deuterium and tritium) combined into a heavier nucleus, release a huge energy with mass loss at the same time, this reaction is called nuclear fusion reaction and the energy released is called fusion energy.

句意：由两个或两个以上氢原子核（如氢的同位素——氘和氚）结合成一个较重的原子核，同时发生质量亏损并释放出巨大能量的反应叫作核聚变反应，其释放出的能量称为核聚变能。

[5] Wind power generation technology involves aerodynamics, automatic control, mechanical transmission, electrical machinery, mechanics and materials science and other multidisciplinary comprehensive high-tech system engineering.

句意：风力发电技术是涉及空气动力学、自动控制、机械传动、电机学、力学和材料学等多学科的综合性高技术系统工程。

[6] The development of biomass energy technology includes biomass gasification technology, biomass curing technology, biomass pyrolysis technology, biomass liquefaction technology, biodiesel technology and biogas technology, etc.

句意：生物质能的开发技术包括生物质气化技术、生物质固化技术、生物质热解技术、生物质液化技术、生物柴油技术和沼气技术等。

[7] The range of hydrogen production technology is very wide, including fossil fuel hydrogen production, electrolytic water hydrogen production, solid polymer electrolyte electrolytic hydrogen production, high temperature water vapor electrolysis hydrogen production, biological hydrogen production, biomass hydrogen production, thermochemical decomposition of water hydrogen production, methanol reforming hydrogen production and H_2S decomposition hydrogen production.

句意：制氢技术范围很广，包括化石燃料制氢、电解水制氢、固体聚合物电解质电解制氢、高温水蒸气电解制氢、生物制氢、生物质制氢、热化学分解水制氢、甲醇重整制氢、H_2S 分解制氢。

[8] The car as a means of transportation discharges a large amount of carbon, nitrogen, and a variety of air pollutants every day, such as sulfur oxides, hydrocarbons and lead compounds, resulting in atmospheric pollution, and serious harm to human health and ecological environment.

句意：作为交通工具的汽车，每天要排放大量的碳/氮/硫的氧化物、碳氢化合物、铅化物等多种大气污染物，是重要的大气污染发生源，对人体健康和生态环境带来严重的危害。

Glossary

aerodynamics *n.* 空气动力学
biodiesel *n.* 生物柴油
biogas *n.* 沼气
biomass *n.* 生物质

biomass curing　生物质固化
coordination hydride hydrogen　配位氢化物
digestion　*n.* 消化
energy-containing resource　含能体能源
energy saving and emission reduction　节能减排
ethanol fuel　燃料乙醇
fossil fuel　化石燃料
fuel cell electric vehicle　燃料电池电动汽车
gasification　*n.* 气化
geothermal energy　地热能
geothermal field　地热田
geothermal hot spot　地热点
hydrocarbon　*n.* 烃，碳氢化合物
hydrogen engine　氢发动机
hybrid electric vehicle　混合电动汽车
hydrogen engine vehicle　氢发动机汽车
H_2S decomposition　硫化氢分解
inverter　*n.* 逆变器
line testing　线路测试
locomotive　*n.* 机车
magma　*n.* 岩浆
marine water　海水
methanol　*n.* 甲醇
metal hydride hydrogen　金属氢化物
methanol reforming　甲醇重整
MH/Ni battery　镍电池
multiple driving system　多驱动系统
non-renewable energy　不可再生能源
nuclear decay　核衰变
nuclear fusion　核聚变
nuclear fission　核裂变
ocean energy　海洋能
oxyhydrogen flame　氢氧焰
photochemical conversion　光化学转换
photoelectric conversion　光电转换
photosynthesis　*n.* 光合作用

photothermal conversion　光热转换
photovoltaic power generation　光伏发电
physical adsorption　物理吸附
plasma　*n.* 等离子体
primary energy　一次能源
propane　*n.* 丙烷
proton exchange membrane　质子交换膜
pure electric vehicle　纯电动汽车
renewable energy　可再生能源
seawater temperature difference energy　海水温差能
seawater salinity difference energy　海水盐度差能
secondary energy　二次能源
shunting locomotive　调车机车
single driving system　单驱动系统
solar cell　太阳能电池
solar energy　太阳能
solar panel　太阳能电池板
solar photothermal technique　太阳能光热技术
spontaneous thermonuclear reaction　自发性热核反应
sulfur oxide　氧化硫
thermochemical decomposition　热化学分解
tidal energy　潮汐能
wind energy　风能
windmill　*n.* 风车
wind turbine　风力发电机

Unit 20　Development of Modern Manufacturing

20.1　Introduction

Although it is difficult to be more precise, manufacturing may date back to about 5000—4000 B.C. It is older than recorded history, because primitive cave or rock markings and drawings were dependent on some form of brush or marker using a "paint", or a means of notching the rock, appropriate tools had to be made for these applications. Manufacturing of products for various uses began with the production of articles made of wood, ceramic, stone, and metal. The materials and processes first used to shape products by casting and hammering have been gradually developed over the centuries, using new materials and more complex operations, at increasing rates of production and at higher levels of quality.

Now manufacturing has been developed into a production business that is intended to make profits through converting raw materials into products. Profits result from the product development, quality, reliability, pricing, public image, productivity, team work, and so on. The primary objective of a manufacturing business is to convert raw materials into quality goods that have value in the marketplace and can be sold at competitive prices. The goods are produced through good management techniques in the use of such resources as capital, human labor, materials, equipment, and energy involving many activities and operations. Manufacturing may be defined as a series of interrelated activities and operations including the design, selection of materials, planning, production, quality assurance, management, and marketing of discrete consumable and durable goods.[1] The interaction among these activities and operations form a total manufacturing system. A system is an organized collection of human resources; machines, tools, and equipment resources; financial resources; and methods required to accomplish a set of specific functions. Many processes are used in meeting the primary objective of a manufacturing system in transforming or converting a set of input parameters of new materials and shapes into an output of proper size, configuration, and performance according the specification for the manufacturing system.[2]

People continue to seek ways to produce devices that will improve their standard of living, meet their basic needs, and control their environment. Such improved methods create demands for larger quantities of devices that are produced more quickly than previously and are of better quality at a lower cost per item.

20.2 Mechanization

The construction and application of simple machines for production started in Europe around 1670. These developments moved production from the home to factories, marking the beginning of the Industrial Revolution. Mechanization signified the movement from making products by hand in the home to making products by machines in factories.

Mechanization also created a system of mass production, which placed a demand on machines to duplicate parts with a high degree of accuracy. This resulted in a need for more accurate measuring tools, improved measuring techniques, and standards to help manufacturers make interchangeable parts. Fixed automation mechanisms and transfer lines were major results of mass production. A transfer line is an organization of manufacturing facilities for faster output and shorter production time.

Fixed automation gave way to machine tools with simple automated controls. This type of controller operates automatically to regulate a controlled variable or system. Advancement in controller technology opened the new era of automation, called programmable automation.

20.3 Programmable Automation

Programmable automation is designed to accommodate changes in a product. A new technique in automation, numerical control (NC), was developed around 1952. Numerical control is based on digital computer principles, a form of programmable automation that controls manufacturing processes by numbers, letters, and symbols. Advances in computer technology extended NC to direct numerical control (DNC), computer numerical control (CNC), graphical numerical control (GNC), and voice numerical control (VNC).[3]

Numerical control caused a revolution in the manufacturing of discrete metal parts. The success of NC led to a number of extensions such as adaptive control (AC) and industrial robots. Adaptive control determines the proper speeds and feeds during machining as a function of variations in such factors as hardness of work material, width or depth of cut, and so on. [4] It denotes a control system that measures certain output process variables and uses these to control the speed and feed of the machine tool. Typical process variables used in AC machining systems are spindle deflection or force, torque, cutting temperature, vibration amplitude, and horsepower. Industrial robots started playing a major role in manufacturing during the late 1970s. Initially, robots were used for material handing. Today's robotic technology, however, has been developed to such an extent that robots are used to perform many high-level tasks in manufacturing.

20.4 Computer-aided Manufacturing

Computers are being given an increasingly important role in manufacturing systems. A computer's ability to receive and handle large amounts of data, coupled with the fast processing speed, makes it become an indispensable approach in modern manufacturing system. The use of computers in manufacturing is now coming of age. Computer application in manufacturing production is typically referred to as computer-aided manufacturing (CAM). It is built on the foundation of such systems as NC, AC, robotics, automated guided vehicle system (AGVS), and flexible manufacturing system (FMS).

Computer-aided manufacturing is the effective use of computer technology in the planning, management, control, and operation of a manufacturing production facility through either direct or indirect computer interface with physical and human resources of the company. Computers play important roles in CAM systems, they integrate manufacturing data into a common database. Database management concepts are applied to CAM operations to speed up data access and to ensure that all users work from a common design. Under the CAM definition, computer applications include such systems as inventory control, scheduling, machine monitoring, and management information. These applications are primarily for transferring, interpreting, and keeping track of manufacturing data.

20.5 Flexibility

Flexibility is an important feature in modern manufacturing. It means that a manufacturing system is versatile and adaptable, while also capable of handling relatively high production runs.[5] A flexible manufacturing system is versatile in that it can produce a variety of parts. It is adaptable in that it can be quickly modified to produce a completely different line of parts. This flexibility can be the difference between success and failure in a competitive international marketplace.

In manufacturing there have always been trade-offs between production rates and flexibility. Transfer lines are capable of high-volume runs, but they are not very flexible. At the other end of the spectrum are independent CNC machines that offer maximum flexibility, but are only capable of low production rates. Flexible manufacturing is an attempt to use technology in such a way as to achieve the optimum balance between flexibility and production runs.

Transfer lines are capable of producing large volumes of parts at high production rates. The line takes a great deal of setup, but can turn out identical parts in large quantities. Its chief shortcoming is that even minor design changes in a part can cause the entire line to be shut down and reconfigured.[6] This is a critical weakness because it means that transfer lines cannot produce different parts, even parts from within the same family, without costly and time-consuming shutdown and reconfiguration.[7]

Traditionally, CNC machines have been used to produce small volumes of parts that differ slightly in design. Such machines are ideal for this purpose because they can be quickly reprogrammed to accommodate minor or even major design changes. However, as independent machines, they cannot produce parts in large volumes or at high production rates.

A flexile manufacturing system (FMS) is a group of CNC machine tools served by an automated materials handling system that is computer controlled and has a tool handling capability.[8] Because of its tool handling capability and computer control, such a system can be continually reconfigured to manufacture a wide variety of parts. This is why it is called a flexible manufacturing system.

An FMS can handle higher volumes and production rates than independent CNC machines. What is particularly significant about the capabilities of flexible manufacturing is that most manufacturing situations require medium production rates to produce medium volumes with enough flexibility to quickly reconfigure to produce another part of product. Flexible manufacturing fills this long-standing void in manufacturing.

Flexible manufacturing represents a major step toward the goal of fully integrated manufacturing in that it involves integration of automated production processes. In flexible manufacturing, the automated manufacturing machine (i.e., lathe, mill, drill) and the automated materials handling system share instantaneous communication via a computer network. This is integration on a small scale.

Flexible manufacturing takes a major step toward the goal of fully integrated manufacturing by integrating several automated manufacturing concepts:

1. Computer numerical control (CNC) of individual machine tools;
2. Distributed numerical control (DNC) of manufacturing systems;
3. Automated materials handing systems;
4. Group technology (families of parts).

When these automated processes, machines, and concepts are brought together in one integrated system, an FMS is the result. Humans and computers play major roles in an FMS. The amount of human labor is much less than with a manually operated manufacturing system. However, humans still play a vital role in the operation of an FMS.

20.6 Remanufacturing

As manufacturing generates mare than 60% of annual non-hazardous waste, increasingly severe legislation demands a reduction in the environmental impacts of products and manufacturing processes. For example, producer responsibility legislation requires producers to recover used products to reduce landfill. Such pressures, combined with severe competition due to global industrial activities, challenge companies to alter attitudes to product design. Companies must design products for longevity and ease of recovery of their materials at end of life, and must consider the business potential of processing used products to harness the residual value in their components.

Remanufacturing, a process of bringing used products to a "like-new" functional state, can be both profitable and less harmful to the environment than conventional manufacturing as it reduces landfill and consumption of virgin material, energy and specialized labor force used in production.[9] Key remanufacturing barriers include consumer acceptance, scarcity of remanufacturing tools and techniques and poor remanufacturability of many current products. These result from a lack of remanufacturing knowledge including ambiguity in its definition. The terms repair, reconditioning and remanufacturing are often used synonymously. Consequently, customers are unsure of the quality of remanufactured products and are wary of purchasing them. Also, designers may lack the knowledge to consider end-of-life issues such as remanufacturing in their work because design has traditionally focused on functionality and cost at the expense of environmental issues. In addition, the problems involved in organizational transformations need investigation such as the interactions between innovations on technical, organizational and social development levels and the timescales in these kinds of transformations. The decision-making processes used in manufacturing management, particularly in changing initiatives or transformations need to be evaluated. Remanufacturing approach represents a fundamental change in the decision-making processes of most manufacturers. Historically, the selection of waste management methods was based on pure economic analyses of the quantifiable and measurable costs, and economic benefits. This approach ignores the very large number of qualitative factors affecting the selection of the appropriate technologies in decision making. Remanufacturing, on the other hand, is a complex, multidisciplinary, and multifunctional activity method to determine potentially large numbers of waste minimization technologies available in the industry, for example, changes in the product, changes in the input materials to the production process, changes in operating practices, and recycling. It requires the coordination of several designs and data-based activities, such as environmental impact analysis, data and database management, and design optimizations. There is now a consensus that decision-making is influenced by unpredictability, risk and uncertainty. For example, can the impacts of certain technologies be accurately calculated? How do you identify and calculate risks in different respects, e.g. economics, society and environment? And when certain practices succeed in one organization, why do they fail in another?

Notes

[1] Manufacturing may be defined as a series of interrelated activities and operations including the design, selection of materials, planning, production, quality assurance, management, and marketing of discrete consumable and durable goods.

句意：制造可以定义为一系列相关的活动与操作，包括设计、选材、计划、生产、质量保障、管理，以及离散型消费品和耐用产品的市场营销等。

[2] Many processes are used in meeting the primary objective of a manufacturing system in transforming or converting a set of input parameters of new materials and shapes into an output of proper size, configuration, and performance according the specification for the manufacturing system.

句意：许多制造工艺都根据规范将一组新的材料和形状参数转换成作为输出量的产品尺寸、结构和性能，从而实现某个制造系统的主要目标。

[3] 名词解释：graphical numerical control (GNC), voice numerical control (VNC)。

数控机床在加工过程中，遇到轮廓较复杂的零件时，用人工编写数控程序要花费大量的时间，且易出错，故采用 CAD/CAM 集成技术编制数控加工程序称为图形数控（graphical numerical control, GNC）。语音识别是让机器通过识别和理解过程把语音信号转变为相应的文本或命令的技术，目前应用于很多领域。语音识别技术在机床控制方面也大有用武之地。语音数控（voice numerical control, VNC）机床可以打破人工、地点和设备的限制，突出人性化、智能化。

[4] Adaptive control determines the proper speeds and feeds during machining as a function of variations in such factors as hardness of work material, width or depth of cut, and so on.

句意：自适应控制将加工过程中的工件材料硬度、切削宽度或深度等因素作为变量，来确定合适的加工速度和进给量。

[5] Flexibility is an important characteristic in the modern manufacturing. It means that a manufacturing system is versatile and adaptable; while also capable of handling relatively high production runs.

句意：在现代制造中，柔性是一个很重要的特性，它意味着一个制造系统具有多种功能，并有很强的适应性，而且还能够应付较大批量的生产过程。

[6] Its chief shortcoming is that even minor design changes in a part can cause the entire line to be shut down and reconfigured.

句意：它的主要缺点是即使零件有很小的设计更改，都会造成整个生产线的关闭和重新装备。

[7] This is a critical weakness because it means that transfer lines cannot produce different parts, even parts from within the same family, without costly and time-consuming shutdown and reconfiguration.

句意：这是一个致命的弱点，意味着如果不对自动化生产线进行代价昂贵和耗费时间的关闭与重新装备，它就无法生产不同的零件，即使该零件属于同一零件族也不行。

[8] A flexile manufacturing system (FMS) is a group of CNC machine tools served by an automated materials handling system that is computer controlled and has a tool handling capability.

句意：柔性制造系统（FMS）由一组计算机数控机床和一个为其服务的由计算机控制的具有装卸（搬运）工具能力的自动化物料搬运系统组成。

[9] Remanufacturing, a process of bringing used products to a "like-new" functional state, can be both profitable and less harmful to the environment than conventional manufacturing as it reduces landfill and consumption of virgin material, energy and specialized labor force used in production.

句意：再制造是这样一个过程，它将用过的产品的功能"整旧如新"，此过程不仅是可以获利的，而且比常规制造过程对环境的危害作用小，因为它能减小垃圾填埋坑的占用面积和在生产过程中原材料、能源及具有专门技术的劳力的消耗量。

Glossary

adaptive control (AC)　自适应控制
ambiguity　*n*. 不明确，含糊
automated guided vehicle system (AGVS)　自动导向搬运车系统
be wary of　对……持谨慎态度
development cycle　开发周期
discrete metal part　离散型金属零件
different line of parts　不同种类的零件
distributed numerical control (DNC)　分布式数控
fixed automation　刚性自动化
flexible manufacturing system　柔性制造系统
harness　*v*. 驾驭，控制；开发
hazardous　*adj*. 危险的，冒险的
high-volume run　大量生产

indispensable　*adj.* 不可缺少的
in excess of　超过，多余
inventory control　库存控制
landfill　*n.* 废渣、垃圾填埋地
long-standing　长期存在的
mechanization　*n.* 机械化
monitoring　*n.* 监视，控制，监测
physical resource　物质资源
precautionary　*adj.* 预先警戒的，预防的，留心的
production rate　生产率
programmable automatic　可编程序自动化的
public image　公众形象
quality goods　优质货物
remanufacturing　*adj.* 再制造
residual　*adj.* 残渣的，剩余的　*n.* 残渣，剩余，余数
scheduling　*n.* 编制作业进度计划，调度
signify　*v.* 表示，意味
time-consuming　耗费时间的，旷日持久的
transfer line　自动化生产线
virgin　*adj.* 初始的，原始的
warranty　*n.* 正当理由，授权，担保，保证

Unit 21 IT and Manufacturing

21.1 Introduction

What is IT? IT (information technology) refers to the technology which uses computer science and modern communication approaches to conduct the acquisition, recognition, transfer, storage , processing, display and distribution of information.[1] Over the past few decades, the extensive use of IT in manufacturing has rendered the communication and information exchange to become much more effective and efficient. The application of IT in manufacturing ranges from simple machining applications to manufacturing planning and control support. From the early years of the introduction of numerical control and all the way to today's better communication and information exchange, collaborative design, machining centers, manufacturing cells and flexible systems, remote and networked manufacturing have been the main advantages of IT.[2]

The computer is bringing manufacturing into the Information Age. A well-developed, competitive world is requiring that manufacturing begins to settle for more, to become itself sophisticated.[3] To meet market competition, for example, a company will have to meet the somewhat conflicting demands for greater product diversification, higher quality, improved productivity, and lower prices.[4] The company that seeks to meet these demands will need a sophisticated tool, one that will allow it to respond quickly to customer needs while getting the most out of its manufacturing resources. The computer is that tool.

21.2 Computer-integrated Manufacturing

Becoming a "superquality, superproductivity" plant requires the integration of an extremely complex system. This can be accomplished only when all elements of manufacturing—design, fabrication and assembly, quality assurance, management, materials handling—are computer integrated.

In product design, for example, interactive computer-aided-design (CAD) systems allow the drawing and analysis tasks to be performed in a fraction of the time previously required and with greater accuracy. And programs for prototype testing and evaluation further speed up the design process.

In manufacturing planning, computer-aided process planning (CAPP) allows for the selection of the optimum process from thousands of possible sequences and schedules.

On the shop floor, distributed intelligence in the form of microprocessors controlled machines, runs automated loading and unloading equipment, and collects data on current shop conditions.

But such isolated revolutions are not enough. What is needed is a totally automated system, linked by common software from front door to back.

"Computer-integrated manufacturing" (CIM) is the term used for the complete automation of the factory, with all processes functioning under computer control and digital information tying them together. CIM encompasses many of the other advanced manufacturing technologies such as CNC, CAD/CAM, robotics, and just-in-time (JIT) delivery. Computer integration provides widely and instantaneously available, accurate information, improving communication between departments, permitting tighter control, and generally enhancing the overall quality and efficiency of the entire system.[5]

With the advent of computer age, manufacturing has developed a full circle. Design has evolved from a manual process using such tools as slide rules, triangles, pencils, scales, and erasers into an automated process known as computer-aided design (CAD).[6] Process planning has evolved from a manual process using planning tables, diagrams, and charts into an automated process known as computer-aided process planning (CAPP). Production has evolved from a manual process involving manually controlled machines into an automated process known as computer-aided manufacturing (CAM). These individual components of manufacturing evolved over the years into separate islands of automation. These islands and other automated components of manufacturing are linked together through computer networks.

Modern manufacturing encompasses all of the activities and processes necessary to convert raw materials into finished products, deliver them to the market, and support them in the field. These activities include the following:

1. Identifying a need for a product;
2. Designing a product to meet the needs;
3. Obtaining the raw materials needed to produce the product;
4. Applying appropriate processes to transform the raw materials into finished products;
5. Transporting products to the market;
6. Maintaining the product to ensure proper performance in the field.

With CIM, not only are the various elements automated, but the islands of automation are all lined together or integrated. Integration means that a system can provide complete and instantaneous sharing of information. In modern manufacturing, integration is accomplished by computers. CIM, then, is the total integration of all components involved in converting raw materials into finished products and getting the products to the market (see Fig.21.1).

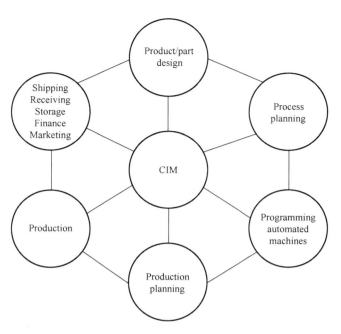

Fig.21.1 Major component of CIM

The future full CIM system contains five main elements: product design, production planning, production control, production equipment, and production processes. Its potential is threefold. First, the computer has unique potential to provide manufacturing with two powerful capabilities, never before available, namely:

1. Online variable program (flexible) automation;
2. Online moment-by-moment optimization.

Second, and very important, the computer has the capability to do the above not only for the hard components of manufacturing (the manufacturing machinery and equipment) but also for the soft components of manufacturing (the information flow, the databases, and so on). Third, and of utmost importance, the computer has the capability to do the above not only for the various links of manufacturing activity but also for the entire system of manufacturing.[7] The computer therefore has tremendous potential to integrate that entire system and thereby produce what is called the computer-integrated manufacturing system.

21.3 Enterprise Resource Planning (ERP) System

The modern enterprise resource planning (ERP) systems incorporate all resource planning and business processes of the entire enterprise, including human resources, project management, product design, materials and capacity planning. The elimination of incorrect information and data redundancy, the standardization of business unit interfaces, the confrontation of global access and security issue, the exact modeling of business processes have all become part of a successful ERP implementation. The ERP systems often provide optimization capabilities, resulting in cost and time savings for variety of manufacturing processes. Indicative examples involve cases from simple optimization problems, shop-floor scheduling and production planning to today's complex decision making problems and real-time production control. Furthermore, the ERP systems often provide both Supply Chain Management (SCM) solutions and interfaces for interacting in an integrated way with other external information technology systems. SCM solutions deal with the current trend of manufacturing companies to maximize their communication and collaboration capacity by integrating their operations with those of their business partners.

21.4 Computer-aided System (CAx)

In combination with the ERP systems, the computer-aided systems(CAx) have boosted productivity, by expanding the creativity of designers, allowing for the early experimentation with novel product and process concepts, minimizing the need for prototypes, taking into consideration the disassembly/recycling process, whilst optimizing energy and material use.[8] Latest CAx solutions integrate design (CAD), process planning (CAPP), engineering (CAE), quality control (CAQC), providing design and styling options, fully reconfigurable process plans, NC code for machining, finite element analysis, kinematic and dynamic analysis, conformance to tolerance specifications and full simulation of geometrical properties, including texture and mechanical characteristics of materials. New approaches for integrating seemingly distant stages, such as process planning and scheduling have already emerged, thus enabling the even more efficient coordination of available resources. Complementarily, collaborative design in digital environments is another emerging and quite promising R&D field. The development of shared virtual environments has enabled dispersed actors to share and visualize data, to interact realistically and make decisions in the context of product and process design activities over the Web. CAx systems help engineering teams resolve problems occurring in one production site. Optimizing the use of materials and energy can be effectively achieved by using these new technologies.

The evolution of ERP and CAx into integrated platforms has led to the new state-of-the-art Product Lifecycle Management (PLM) systems, which allow for the performance of a variety of data management tasks, including vaulting, workflow, lifecycle, product structure, and view and change management. [9] On the other hand, the Product Data Management (PDM) systems are claimed to be able to integrate and manage a number of applications, information types, and process that defines a product, from design to manufacture to end-user support. The ability of manufacturing organizations to integrate business functions and departments with new systems into an enterprise database allows them to have a unified enterprise view. These systems are based on the Digital Factory/Manufacturing concept, according to which production data management systems and simulation technologies are jointly used for optimizing manufacturing before starting production. In principle, by exploiting Digital Manufacturing, manufacturing enterprises are expected to achieve:

1. Shortened product development;
2. Early validation of manufacturing processes;
3. Faster production ramp up;
4. Faster time-to-market;
5. Reduced manufacturing costs;
6. Improved product quality;
7. Enhanced product knowledge dissemination;
8. Reduction in errors;
9. Increase in flexibility.

In addition to the hard core engineering advantages, there are also economic ones, as the competition is increased through the information society (e.g. the Internet). It can therefore be assumed that the positive impact of these new technologies can make up for the enormous material and energy demand. In any case, and in order to produce more with less, flexibility and automation in production are and will be of increasing importance.

Notes

[1] IT (information technology) refers to the technology which uses computer science and modern communication approaches to conduct the acquisition, recognition, transfer, storage, processing, display and distribution of information.

句意：IT（信息技术）是指利用电子计算机和现代通信手段，获取、识别、传递、存储、处理、显示和分配信息的技术。

[2] From the early years of the introduction of numerical control and all the way to today's better communication and information exchange, collaborative design, machining centers, manufacturing cells and flexible systems, remote and networked manufacturing have been the main advantages of IT.

句意：从数控技术的诞生直到今天更加完善的通信与信息交换，信息技术给我们带来的好处有协同设计、加工中心、制造单元和柔性制造、远程与网络化制造等。

[3] A well-developed, competitive world is requiring that manufacturing begins to settle for more, to become itself sophisticated.

句意：一个高度发展的、充满竞争的世界正在要求制造业开始满足更多的需求，并使其自身采用先进高端的技术进行装备。

[4] To meet competition, for example, a company will have to meet the somewhat conflicting demands for greater product diversification, higher quality, improved productivity, and lower prices.

句意：例如，为了适应竞争，一家公司会满足某些多少有点相互矛盾的需求：既要生产多样化和高质量的产品，又要提高生产率和降低价格。

[5] Computer integration provides widely and instantaneously available, accurate information, improving communication between departments, permitting tighter control, and generally enhancing the overall quality and efficiency of the entire system.

句意：计算机集成可提供广泛、及时和精确的信息，可以改进各部门之间的交流与沟通，实施更严格的控制，从而提高整个系统的整体质量与效率。

[6] Design has evolved from a manual process using such tools as slide rules, triangles, pencils, scales, and erasers into an automated process known as computer-aided design (CAD).

句意：设计已经从使用诸如计算尺、三角板、铅笔、比例尺和橡皮擦的手工工艺演变成为一种称为计算机辅助设计（CAD）的自动化工艺。

[7] Third, and of utmost importance, the computer has the capability to do the above not only for the various links of manufacturing activity but also for the entire system of manufacturing.

句意：第三点，也是最重要的，不仅在制造活动的各个环节中，而且在整个制造系统中，计算机都有能力完成上述工作。

[8] In combination with the ERP systems, the computer aided systems (CAx) have boosted productivity, by expanding the creativity of designers, allowing for the early experimentation with novel product and process concepts, minimizing the need for prototypes, taking into consideration the disassembly/recycling process, whilst optimizing energy and material use.

句意：由于和ERP系统相结合，计算机辅助系统通过拓展设计者的创造性，对新产品和工艺过程进行早期试验，最小化对样机的需求，同时考虑拆卸和回收方法，并对能源和材料优化使用，从而大大提高了生产率。

[9] The evolution of ERP and CAx into integrated platforms has led to the new state-of-the-art product lifecycle management (PLM) systems, which allow for the performance of a variety of data management tasks, including vaulting, workflow, lifecycle, product structure, and view and change management.

句意：ERP和CAx的集成化发展形成了一个达到最新技术发展水平的产品寿命周期

管理系统（PLM），从该系统可以得到各种各样的数据管理任务，包括介质管理、工作流程、寿命周期、产品结构，以及检查和改变管理任务。

Glossary

accessibility　　*n.* 可接近性，易维护性，检查，操作
chart　　*n.* 图表
computer-integrated manufacturing　　计算机集成制造
conceive　　*v.* 设想，想象，表现
conflicting　　*adj.* 不一致的，冲突的，矛盾的，不相容的
delivery　　*n.* 交货
disassembly　　*n.* 拆卸，分解
dissemination　　*n.* 散播，传播，普及
diversification　　*n.* 多样化，变化，不同，多种经营
elimination　　*n.* 除去，消除
enterprise resource planning (ERP)　　企业资源计划
fade away　　渐渐消失
flexibility　　*n.* 弹性，柔性
interactive　　*adj.* 交互式的，人机对话交互式的
interchangeability　　*n.* 可交换性，互换性，可替代性
ironic　　*adj.* 讽刺的，令人啼笑皆非的
just-in-time　　即时
link　　*n.* 环节
optimization　　*n.* 最优化
potential　　*adj.* 潜能，潜力
prototype　　*n.* 原型，样件，样机
reflex　　*n.* 反射，映射，回复
schedule　　*n.* 时间表，进度表，计划表，进程，预定计划
shop floor　　车间现场
state-of-the-art　　最先进的，当前最高水平的
threefold　　*adv.* 三倍地，三方面地
tremendous　　*adj.* 极大的，巨大的
triangle　　*n.* 三角形，三角板
workflow　　*n.* 工作流程

Unit 22   Manufacturing and Roles of Engineers

22.1   Manufacturing

In its broadest sense, manufacturing is the process of converting raw materials into products. It encompasses (1) the design of the product, (2) the selection of raw materials, and (3) the sequence of processes through which the product will be manufactured.

Manufacturing is the backbone of any industrialized nation. Its importance is emphasized by the fact that, as an economic activity, it comprises approximately 20% to 30% of the value of all goods and services produced. A country's level of manufacturing activity is directly related to its economic strength. Generally, the higher the level of manufacturing activity in a country, the higher the standard of living of its people will be.

Manufacturing also involves activities in which the manufactured product is itself used to make other products. Examples of these products are large presses to shape sheet metal for car bodies, machinery to make bolts and nuts, and sewing machines for making clothing. An equally important aspect of manufacturing activities is the servicing and maintenance of this machinery during its useful life.

The word *manufacturing* is derived from the Latin *manu factus*, meaning made by hand. The word manufacture first appeared in 1467, and the word manufacturing appeared in 1583. In the modern sense, manufacturing involves making products from raw materials by means of various processes, machinery, and operations, through a well-organized plan for each activity required.[1] The word *product* means something that is produced, and the words product and production first appeared sometime during the 14th century.

The word *production* is often used interchangeably with the word manufacturing. Whereas **manufacturing engineering** is the term used widely in the United States to describe this area of industrial activity, the equivalent term in other countries is **production engineering**.[2] Because a manufactured item has undergone a number of processes in which pieces of raw material have been turned into a useful product, it has a **value**—defined as monetary worth or marketable price. For example, as the raw material for ceramics, clay has a certain value as mined. When the clay is used to make a ceramic cutting tool or electrical insulator, value is added to the clay. Similarly, a wire coat hanger or a nail has a value over and above the cost of the piece of wire from which it is made. Thus manufacturing has the important function of *adding value*.

Manufacturing may produce **discrete products**, meaning individual parts, or

continuous products. Nails, gears, balls for bearings, beverage cans, and engine blocks are examples of discrete parts, even though they are mass produced at high production rates. On the other hand, a spool of wire, a sheet of metal or plastic, and lengths of tubing, hose, and pipe are continuous products, which may be cut into individual pieces and thus become discrete parts.[3]

Manufacturing is generally a complex activity involving a wide variety of resources and activities, such as the following:

- Product design
- Materials
- Machinery and tooling
- Process planning
- Purchasing
- Manufacturing
- Production control
- Shipping
- Marketing
- Sales
- Support services
- Customer services

Manufacturing activities must be responsive to several demands and trends:

1. A product must fully meet design requirements and product specifications and standards.

2. A product must be manufactured by the most *environmentally friendly* and *economical* methods.

3. *Quality* must be *built* into the product at each stage, from design to assembly, rather than tested after the product is made. Furthermore, the level of quality should be appropriate to the product's use.

4. In a highly competitive environment, production methods must be *flexible* enough to respond to changes in market demands, types of products, production rates, production quantities, and in-time delivery requirements.[4]

5. New developments in materials, production methods, and computer integration of both technological and managerial activities in a manufacturing organization must constantly be evaluated with a view to their appropriate, timely, and economical implementation.

6. Manufacturing activities must be viewed as a large system, the parts of which are interrelated. Such systems can now be modeled, in order to study the effect of factors such as changes in market demands, product design, and materials. Various other factors and production methods affect product quality and cost.

7. A manufacturing organization must constantly strive for higher levels of *quality* and *productivity* (defined as the optimum use of all its resources: materials, machines, energy, capital, labor, and technology). Output per employee per hour in all phases

must be maximized. Zero-based part rejection (and consequent reduction of waste) is also an integral aspect of productivity.

Until the Industrial Revolution, which began in England in the 1650s, goods had been produced in batches, with much reliance on manual labor in all aspects of production. Modern mechanization began in England and Europe with the development of textile machinery and machine tools for cutting metals. This technology soon moved to the United States, where it was developed further, including the important advance of designing, making, and using *interchangeable parts*. Prior to the introduction of interchangeable parts, a great deal of hand-fitting was necessary, because no two parts were made exactly alike. By contrast, we now take for granted that we can replace a broken bolt of a certain size with an identical one purchased years later from a local hardware store.[5] Further developments soon followed, resulting in numerous products that we cannot imagine being without, because they are so common. Since the early 1940s, major milestones have been reached in all aspects of manufacturing. For example, great progress was made during the past 100 years, and especially during the last two decades with the advent of the computer age, as compared to that during the long period from 4000 B.C. to 1 B.C.

Although the Romans had factories for mass-producing glassware, manufacturing methods were at first very primitive and generally very slow, with much manpower involved in handling parts and running the machinery. Today, with the help of computer-integrated manufacturing systems, production methods have been advanced to such an extent that, for example, holes in sheet metal are punched at rates of 800 per minute and aluminum beverage cans are manufactured at rates of 500 per minute.

22.2 Roles of Engineers

Mechanical engineers research, develop, design, manufacture and test tools, engines, machines, and other mechanical devices. They work on power-producing machines such as electricity-producing generators, internal combustion engines, steam and gas turbines, and jet and rocket engines. They also develop power-using machines such as refrigeration and air-conditioning equipment, robots used in manufacturing, machine tools, materials handling systems, and industrial production equipment. Mechanical engineers are known for working on a wide range of products and machines. Some examples are automobiles, aircraft, jet engines, computer hard drives, microelectromechanical acceleration sensors (used in automobile air bags), heating and ventilation systems, heavy construction equipment, cell phones, artificial hip implants, robotic manufacturing systems, replacement heart valves, planetary exploration and communications spacecraft, deep-sea research vessels, and equipment for detecting explosives.[6] Being involved in nearly every stage of a product's life cycle from concept through final production, engineers often work as designers in specifying components, dimensions, materials, and machining processes. A mechanical engineer who specializes in manufacturing is concerned with the production of hardware on a day-to-day basis and with assuring consistent quality. A research and development engineer, on the other hand, works over a longer time frame and is responsible for demonstrating new products and technologies. Engineering managers, for instance, organize complex technical operations and help to identify new customers, markets, and products for their company.

Mechanical engineering is the second-largest field among the traditional engineering disciplines and is perhaps the most general. In 1998, approximately 220000 people were employed as mechanical engineers in the United States, representing 14% of all engineers. The discipline is closely related to the technical areas of industrial (126000), aerospace (53000), petroleum (12000), and nuclear (12000) engineering, each of which evolved historically as a branch or spin-off of mechanical engineering. Together, mechanical, industrial, and aerospace engineering account for about 28 % of all engineers. Other specializations that are frequently encountered in mechanical engineering include automotive, design, and manufacturing engineering. Mechanical engineering is often regarded as the broadest and most flexible of the traditional engineering disciplines, but there are many opportunities for specialization within a certain industry or technology that is of particular interest. Within the aviation industry, for example, an engineer will further focus on a single core technology, perhaps jet engine propulsion or flight control systems. An engineer's contribution is ultimately judged based on whether the product functions as it should.[7]

Mechanical engineering is driven by the desire to advance society through technology. The American Society of Mechanical Engineers (ASME) surveyed its members at the turn of the millennium in order to identify mechanical engineering's major achievements. This "top-ten" list, summarized here, known as the milestones of modern manufacturing, includes: (1) Automobile, (2) Apollo program, (3) Power generation, (4) Agricultural mechanization, (5) Airplane, (6) Integrated circuit mass production, (7) Air-conditioning and refrigeration, (8) Computer-aided engineering technology, (9) Bioengineering, (10) Codes and standards.[8]

Notes

[1] In the modern sense, manufacturing involves making products from raw materials by means of various processes, machinery, and operations, through a well-organized plan for each activity required.

句意：从现代的观点来看，制造是指借助各种工艺手段、机器设备与操作，并通过组织完善的生产计划完成所需的每一步生产活动，从而将原材料制成产品的全部过程。

[2] The word production is often used interchangeably with the word manufacturing. Whereas manufacturing engineering is the term used widely in the United States to describe this area of industrial activity, the equivalent term in other countries is production engineering.

句意："生产"一词常常和"制造"交换使用，而"制造工程"这一术语在美国则被经常用来描述工业活动，其他国家采用的等效词汇为"生产工程"。

[3] Manufacturing may produce discrete products, meaning individual parts, or continuous products. Nails, gears, balls for bearings, beverage cans, and engine blocks are examples of discrete parts, even though they are mass produced at high production rates. On the other hand, a spool of wire, a sheet of metal or plastic, and lengths of tubing, hose, and pipe are continuous products, which may be cut into individual pieces and thus become discrete parts.

句意：制造可以生产离散型产品，也就是单个的产品，也能生产连续型产品。像钉子、齿轮、轴承滚珠、饮料罐头及发动机缸体等就属于前者，即使它们是以大规模方式生产出来的；而钢丝卷材、金属或塑料板材、管材、软管及管子等（连续型产品）也可以切割成离散件。

[4] In a highly competitive environment, production methods must be flexible enough to respond to changes in market demands, types of products, production rates, production quantities, and in-time delivery requirements.

句意：在一个竞争激烈的环境下，生产方法必须有足够的柔性以应对市场需求、产品类型、生产率、生产量及准时供货要求等因素的变化。

[5] Prior to the introduction of interchangeable parts, a great deal of hand-fitting was necessary, because no two parts were made exactly alike. By contrast, we now take for granted that we can replace a broken bolt of a certain size with an identical one purchased years later from a local hardware store.

句意：在可互换性零件出现以前，我们需要大量的手工装配，因为没有任何两个零件会制造得完全一样。而相比之下，如果现在我们发现一个螺栓使用几年以后损坏了，要从当地一家五金店购置和它完全一样的零件来替换，是完全不成问题的。

[6] Mechanical engineers are known for working on a wide range of products and machines. Some examples are automobiles, aircraft, jet engines, computer hard drives, microelectromechanical acceleration sensors (used in automobile air bags), heating and ventilation systems, heavy construction equipment, cell phones, artificial hip implants, robotic manufacturing systems, replacement heart valves, planetary exploration and communications spacecraft, deep-sea research vessels, and equipment for detecting explosives.

句意：机械工程师要和各种产品与机器打交道，涉及汽车、飞机和喷气式发动机，计算机硬盘，微机械电子加速度传感器（用于汽车气囊），加热与通风系统，重型建筑设备，手机，人工髋关节植入体，机器人制造系统，人工替代心脏瓣膜，星球探测与通信飞船，深海探测船，以及爆炸物探测装置等。

[7] Within the aviation industry, for example, an engineer will further focus on a single core technology, perhaps jet engine propulsion or flight control systems. An engineer's contribution is ultimately judged based on whether the product functions as it should.

句意：例如，在航空工业中，工程师会进一步关注某一核心技术，也许是喷气式发动机的推进与飞行控制系统。工程师的贡献最终要由他们所设计与制造的产品能否正常发挥其应有的功能来评价。

[8] This "top-ten" list, summarized here, known as the milestones of modern manufacturing, includes: (1) Automobile, (2) Apollo program, (3) Power generation, (4) Agricultural mechanization, (5) Airplane, (6) Integrated circuit mass production, (7) Air-conditioning and refrigeration, (8) Computer-aided engineering technology, (9) Bioengineering, (10) Codes and standards.

句意：这里总结的被誉为现代制造技术里程碑的机械工程"十大"成就包括：（1）汽车，（2）阿波罗计划，（3）发电，（4）农业机械化，（5）飞机，（6）集成电路大规模生产，（7）空调与制冷技术，（8）计算机辅助工程（CAE），（9）生物工程，（10）（制造）规范与标准。

Glossary

acceleration sensor　加速度传感器
adding value　增值
artificial hip implant　人工髋关节植入体
ASME　美国机械工程师学会
backbone　*n.* 骨干
cannot imagine being without　不能想象没有
cell phone　手机，移动电话
deep-sea research vessel　深海探测船
discipline　*n.* 学科
discrete products　离散型产品
engine block　发动机缸体
generator　*n.* 发电机
heart valves　心脏瓣膜

hose *n.* 软管，水龙带
Industrial Revolution 工业革命
interchangeable parts 可互换性零件
internal combustion engine 内燃机
monetary worth 货币价值
part rejection rate 零件报废率
propulsion *n.* 推进
replacement heart valve 人工替代心脏瓣膜
spin-off 派生，衍生
spool *n.* 卷
strive for 争取，奋斗
turbine *n.* 透平机，涡轮机

附录 A 科技论文英文摘要写作要点

A.1 论文标题

1. 标题不是句子，而是名词词组，不要求主、谓、宾齐全。
2. 标题中尽量避免 study on、research on、investigation of 之类没有多少实际意义的词组，尤其是在题目较长的情况下。
3. 标题中尽量不要用缩略语（除非是大家熟知的）、化学式和专利商标名。
4. 第一个词和实词的首字母要大写（其他的词如 the、a、an、and 和所有介词都小写）。
5. 缩略词的首字母如果是 A、E、F、H、I、L、M、N、O、R、S、X，则前面的不定冠词用 an（因为这 12 个字母发元音）。缩略词的复数在后面加 s 或 es 要根据最后一个字母确定，如 ICs、FMSes。

A.2 摘要写作注意事项

1. 摘要的基本内容为论文的研究目的（purpose）、主要研究过程（procedure）、采用的方法（methods）及主要结论（conclusion）。
2. 摘要写作时，尽可能用短句、被动语态和第三人称来写。
3. 描述论文背景、理由和研究目标等用一般现在时，描述作者以往的科研工作及实验可用过去时或现在完成时，但一般真理、定理、公式和结论用一般现在时。
4. 不能在毫无把握的情况下想当然造字，专业词汇一定要准确，应以英、美学者写的原文为参考标准。查找英语技术词汇不建议使用金山词霸。目前，比较实用的纸质字典有《英汉科学技术词典》（国防工业出版社）、《新英汉机械工程词汇》（科学出版社）；电子词典有联网的 CNKI 翻译助手。
5. 尽量不连用三个以上的 of（可用所属格'、for 或从句表示）。
6. 单复数主语应有对应的谓语动词形式，这种类型的错误时常可见，如 data 是复数形式（单数形式是 datum。注意：它又是加工、测量基准的意思），谓语动词要用复数形式；而 games（大型运动会）、proceedings（论文集）、news、United States、United Nations、airlines（航空公司）、headquarters（总部）看起来是复数形式，实际上是单数名词。
7. 论文种类有：thesis 是本科生或硕士研究生毕业论文，dissertation 是博士生毕业论文；而一般在学术杂志上刊登的论文称为 paper 或 article。千万不能混用。
论文性质有：应用基础研究（应用型工科学生的毕业论文）是 applied basic research，

基础研究（理科学生的论文）是 basic research。

8. 要用多个同义词代替某个词，不要总是重复使用一个词，显得文章乏味，缺少文采。例如：（以下只给出科技文章中常用动词）

（1）做、开展、进行、从事

do, make, carry out, perform, conduct (research, study, experiment, survey, investigation, exploration, cooperation, …)

（2）制造、制作

make, fabricate, build, manufacture, construct, …

（3）改变、改换、修改、重建、重组

convert, transform, change, modify, adapt, alter, reform, correct, reconstruct, rebuild, reconfigure, reorganize, …

（4）建议、提出

advise, recommend, suggest, propose, introduce, put forward, present, develop, …

（5）建立

establish, set up, form, construct, formulate, build, …

（6）给、提供

give, offer, provide, supply, furnish

（7）改善、改进

improve, enhance, raise, better

（8）开发、利用

develop, exploit, take advantage of, make use of, employ, tap, open up, …

（9）验证、证明

verify, prove, check, test, identify, justify, affirm, confirm

（10）占有、构成

constitute, make up, account for, cover, hold, occupy

9. 科技文章中一些常用词意义上的差别

（1）manufacture、fabricate 和 make 的区别

manufacture 一般指工业上批量的、成熟的、有一定规模的制造；而 fabricate 一般指具有技巧性手工方式的制造，而且一般是小批量甚至单件的制作；make 泛指做和制造。

（2）transform 指根本的转换、变换、改造（常跟 into）；convert 只指物理形式的转换（如数模转换）；transfer 和 transmit 指传递、传送、发射等（如 heat transfer 传热学、data transmission 数据传输、power transmission 功率传递、TV transmission tower 电视发射塔、hydraulic transmission 液压传动），特别值得注意的是，technological transformation 是技术改造，但 technology transfer 却是技术转让，transformation matrix 是转换矩阵；communication 指信息交流、通信、联络；transport 指交通运输；transit 指转运、转口、过渡（如 phase transition 是相变、during the transition of the centuries 是世纪之交、trans-century talent 是跨世纪人才）。

（3）enterprise management 是企业管理，project management 是项目管理；而 business administration 是工商管理，public administration 是公共管理，data administration 是数据管理，administrative operator 是管理操作员。

（4）tool 是工具，cutter 是刀具，tooling（无复数形式）是工装、模具，device 是小装置，equipment（无复数形式）是设备，instrument 是仪器仪表、乐器（注意，instrument 还有手段、方法、证书等意思），installation 是成套装置设备，apparatus 是电器、机电设备，appliance 是家用（电）器具，utensil 是器皿厨具，facility（常用复数 facilities）是设施、设备、工具、机构（如 sport facilities 体育器材、设施），utility 是公共事业设备（水、电、煤气等），implement 是工具、器械（农具）。

（5）element、part 是零件，component 是组件、元件、组分、分量，subassembly 是组合件、机组、分组成、分部装配，assembly 是总装、装配，machine 是机器，machine tool 是机床，machinery 是机械总称。

（6）accuracy 是精度（如 machining accuracy 是加工精度，locating accuracy 是定位精度，positional accuracy 是位置精度，measuring accuracy 是测量精度）；precision 是精密（如 precision instrument 是精密仪器，precision machining 是精密加工，precision mold 是精密模具）。

（7）corrosion、erosion、etching、ablation、abrade、rust 的区别
corrosion 特指有酸碱作用下的化学腐蚀；erosion 一般指风蚀、冲蚀，是一种在物理、化学、机械及气候变化条件下对材料的破坏腐蚀作用；etching 指人为的腐蚀或刻蚀（如磨金相试样及制作印制电路板）；ablation 是在热作用下的烧蚀；abrade 指磨蚀、磨损；rust 专指锈蚀（由氧化作用造成）。

（8）location 和 position 的区别
location 只作名词，意思是位置或定位，但这个定位是指将某个物体放在或固定在空间某个具体位置，如 six-point location principle 是六点定位原则，locating pin 是定位销，locating accuracy 是定位精度；而 position 既是名词又是动词，指确定或找到某个物体的确切位置，如 GPS（Global Positioning System）是全球定位系统，其中 positioning 显然是动词 position 衍生成的分词形式。另外，position 还有一个形容词形式 positional，如 positional tolerance 是位置公差，positional distance 是位置距离。

（9）transducer 和 sensor 的区别
在英文文献中常常遇到 transducer 和 sensor 的混用。严格来说，transducer 称为换能

器，它是将输入能量转换成另一种形式的能量的装置。而且，它在不同专业领域还有变送器（如闭环反馈装置中）和变频器（如电梯与空调的调速变频控制中）等名称。而 sensor 称为传感器、感测器，它将物理、化学、机械，光学等过程中可测量量转换成对技术人员或计算机有意义的数据。一般来说，transducer 需要有转换（放大）电路，而 sensor 有可能不需要这样的电路。例如，压电晶体、光电管属于 transducer，而普通温度计应该算是 sensor。还有一个词 probe，俗称测头、探测器、探头，一般是指简单的传感器，或者 transducer 及 sensor 的口语或简单表达词。另外，sensing element 是指敏感元件，如光敏电阻、热敏电阻等。

（10）intensity 和 strength 的区别

intensity 指强化的程度、力度（如 increase the investment intensity 是加大投资力度，strengthen the training intensity 是提高训练强度）；strength 指材料受力后能承受而不被破坏的能力，如力学强度（如 fatigue strength 是疲劳强度，ultimate strength 是极限强度）；另外一个特殊的用法是，strength 有时也指优势和实力，如 on the strength of 是凭借……方面的实力。

（11）base、basis 和 basic(al)的区别

base 指基地、机座（如 experimental base 是实验基地，incubation base 是孵化基地）；basis 指（抽象的）基础（如 on the basis of 是在……基础上）；basic(al)则是基本的含义。

（12）complete、accomplish 和 finish 的区别

complete 用于完成某个项目、工程（project、engineering）；accomplish 用于完成某项目标、指标（goal、index）；finish 则用来完成作业（homework），学业（……with school），论文（thesis、dissertation……）等。另外，complete 还是形容词，表示完整的、全套的、整个的等，如 complete equipment 是全套设备、complete spare parts 是成套备件。

（13）need、demand、requirement(s)、request 和 solicit 的区别

need 是指（人类）自然的需要、急需；demand (for)是指市场、物质需求（如 to meet the market demand 是满足市场需求）；requirement(s)是指对……提出的要求；request 是指祈求、强烈请求、恳求（如 make a request for 是提出请求）；solicit 是指（只有动词）恳求、祈求、要求、征求，后跟 for 或 to + inf.，如 solicit assistance 是请求支援。

（14）schematic diagram、schematic illustration 是示意图，block diagram 是方框图，flow chart 是流程图，demonstration figure 是演示图。

（15）relation 和 relationship 的区别

relation 指人际、国际、贸易关系，而科学研究、实验中的各影响因素之间的关

系都用 relationship (between, among, of, with)。例如：

① 实验分析了流量与压力、温度和流道结构等其他影响因素之间的关系。

The experiment has analyzed the relationship between flow rate and such affecting factors as pressure, temperature and channel structure, etc.

② 该模拟计算探讨了 A、B、C、D 四者之间的相互（交错）关系。

This simulation discussed the relationship between and among A, B, C and D.

另外，关于 relative、related、relevant 的区别有：relative to 是相对……的，related to 是与……有关的，而 relevant 是相关的。

A.3 典型常用语句实例

1. 本文提出了一种基于……的算法，用于改善……的收敛速度与计算精度。

    The paper presents a …-based algorithm to improve the convergence speed and calculation accuracy.

2. 本文开发了一种……装置，可大大提高……系统的稳定性与抗干扰能力。

    The paper developed a … device, this device could greatly enhance (raise, improve) the stability and anti-jamming ability of … system.

3. 本文发明了一种新型……装置，解决了传统……系统加工精度低、成本高、产品性能不稳定的问题。

    The paper invented a novel…device, solved such problems as low accuracy, high cost and unstable product performance when using traditional…system to machine parts.

4. 从实验结果可以看出，当……值升高的时候，……也将随之升高，而……值却呈下降趋势。至于……值的少许波动，作者认为可能是…造成的结果。如果溶液浓度或电流密度能控制稳定的话，这种现象就会得到缓解甚至消除。……等人在关于……的一篇科学报告中已有解释。

    It can be seen from the experimental results that when … increases, … will also increase and … shows a decreasing tendency. As for the slight fluctuation of …, the author thinks it may be caused by … . If the solution concentration or current density could be controlled stable, this phenomenon may acquire alleviation even elimination. On this problem, … et al. have explained in a scientific report about …

5. X 与 3 个变量 Y、Z 和 W 之间的关系可以从图 4 中清楚地显示出来。

    The relationships of X with Y, Z and W can be shown clearly from Fig.4.

6. 从该特性曲线可以看出，温度、压力与……浓度对测试结果影响很大。

    It can be seen from the characteristic curve that temperature, pressure and the concentration of ... have great effects influence on the measured results.

7. 我们利用两相流理论分析了微流道内杂质颗粒的运动速度和浓度分布情况，从而为研究该新型流道的抗堵性能提供了理论依据。

   We used two-phase flow theory to analyze the motion velocity and concentration distribution of the impurity particles inside the micro channels and hence provided a theoretical basis for the study on the anti-clogging behavior of new type emitters.

8. 实验结果表明，该装置的各项电气参数均符合国家……标准，而且耐候性能大大优于进口同类产品，填补了我国在……的空白。

   The experimental results show that various electrical parameters of this device can meet the national standard of ... and its weather resistance is much better than the imported same products, which filled in the blank of ... in our country.

9. 通过实际切削加工，证明本文所开发的……新型装置可以明显地提高加工件的形状精度和表面光洁度，而且生产成本也大大降低，为……产品打入国际市场提供了优良的设备保障。

   Through practical cutting process, it is proved that the ... novel device developed in this thesis can obviously raise the shape accuracy and surface finish of the workpieces and the production cost is also greatly reduced, providing an excellent equipment guarantee for putting ... into international market.

10. 这种多孔夹层复合材料的优良综合性能使得它在航空航天、运动器材和国防工业上得到广泛应用。

    This porous sandwich composite's comprehensive properties make it find extensive applications in aerospace industry, spots facilities and defense industry.

A.4 汉译英范文（参考）

进入21世纪，IT行业的飞速发展使得中国在IT行业中拥有巨大的市场。IT产品中的关键部件是集成电路芯片，其附加值最大。比如，我国在移动通信制造业中，因为缺乏IC的开发和生产能力，尽管市场繁荣，却无利润甚至亏损，其原因就在于此；还有在数字化电视、数控机床、计算机、智能卡、智能电器等众多IT产品中，IC都是核心部件。在现代化的武器装备中，控制用的计算机及其芯片均是自己生产的。IC的设计制造水平及其快速开发能力是占领军事技术制高点的关键。日本IT行业的发展得益于其IC装备制造业的发展。因此，在"十一五"期间发展IC制造装备是一项具有重大战略意义的决策。

在这方面我国的实际情况是，技术水平落后国际水平5代之遥。由于IT行业的高额利润和高度竞争性，我国的芯片制造业的关键设备主要依赖进口，即使花费高额代价也只能得到国外10年前的IC制造装备，这导致我国的IT行业及其他数字化产品乃至军工高端产品都落后于国外 2—3 代。当前，我国台湾地区的 IC 业正在向大陆转移，国内在 IC 装备方面将具有数百亿美元的市场空间。

要占据国内的巨大市场，实现 IC 装备的自主化，技术策略的选择非常关键。要明确的是，要占领市场的技术优势不是相对地缩小与主流技术之间的差距，而是要追赶并超过世界主流技术。所以，只有开发创新的原理和工艺，才有可能实现我国 IT 行业的跨越式发展。

The rapid development of IT industry provides China with a huge market in the 21st century. However, the core component of IT industry is the integrated circuit (short for IC) chip because of its very high added value. Take mobile communication for example, it is because of the poor IC development capability in our country that many businesses rarely make any profit even suffer heavy losses in spite of the prosperous IT market. IC is not only the critical component of communication facilities, but also the heart of digital video television, numerical control, computer, intelligent card, intelligent appliance and numerous other IT products. All computers and ICs in modern military facilities have to be produced domestically, and thus IC design and manufacturing level and the rapid development capability have increasingly become pivotal to the military technology. As is well known that, the advanced IT industry in Japan benefited tremendously from its highly developed IC technology. In view of this, it will be a momentous decision for us to devote major efforts to developing our IC industry during the 11th Five-Year Plan.

The developed countries have left us behind as far as 5 generations in IC industry. Because of the high profit and fierce competition, the key IC manufacturing equipment in our country are mainly imported from abroad. And even at a high cost, what we actually get, however, are nothing but those used 10 years ago overseas, which results in such a reality that our IT industry, digital products as well as some sophisticated military equipment fall behind those of the developed countries by 2—3 generations. Presently, Taiwan IC industry is directing toward the mainland market whose potential is worth tens of billions of US dollars.

It is extremely urgent to make a strategic decision to seize the tremendous Chinese market and eventually realize the independent and self-reliant development of our IC equipment. In addition, we must understand that what we should do is keeping up with the advanced mainstream IC technology rather than just narrowing the gap in between. Therefore, only through developing innovative principles and technologies can we realize a great-leap-forward development of China's IT industry.

附录 B 麻省理工学院简介

B.1 原文

Massachusetts Institute of Technology

The Massachusetts Institute of Technology (MIT) is a private research university located in Cambridge, Massachusetts, United States. MIT has five schools and one college, containing a total of 32 academic departments, with a strong emphasis on scientific and technological research. MIT was the first university in the nation to have a curriculum in: architecture, electrical engineering, sanitary engineering, naval architecture and marine engineering, aeronautical engineering, meteorology, nuclear physics, and artificial intelligence.

The four-year, full-time undergraduate instructional program is classified as "balanced arts & sciences/professions". The graduate program is classified as "comprehensive". The university is accredited by the New England Association of Schools and Colleges.

Several rankings place MIT among the top colleges and universities in the United States and internationally. The School of Engineering has been ranked first among graduate and undergraduate programs by U.S. News and World Report since first published results in 1994. A 1995 National Research Council study of US research universities ranked MIT first in "reputation" and fourth in "citations and faculty awards" and a 2005 study found MIT to be the 4th most preferred college among undergraduate applicants.

The mission of MIT is to advance knowledge and educate students in science, technology, and other areas of scholarship that will best serve the nation and the world in the 21st century.

The Institute admitted its first students in 1865, four years after the approval of its founding charter. The opening marked the culmination of an extended effort by William Barton Rogers, the founder of MIT and a distinguished natural scientist, to establish a new kind of independent educational institution relevant to an increasingly industrialized America. Rogers believed that professional competence is best fostered by coupling teaching and research and by focusing attention on real-world problems.

Today MIT is a world-class educational institution. Teaching and research—with relevance to the practical world as a guiding principle—continue to be its primary purpose. MIT is independent, coeducational, and privately endowed. Its five schools and one college encompass numerous academic departments, divisions, and degree-granting programs, as well as interdisciplinary centers, laboratories, and programs whose work cuts across traditional departmental boundaries.

The Campus

MIT is located on 158 acres that extend more than a mile along the Cambridge side of the Charles River Basin. The central group of interconnecting buildings, dedicated in 1915, was designed to permit easy communication among schools and departments. Subsequent growth of the campus saw construction of landmark buildings by some famous leading architects.

Currently under construction are three new academic and research buildings. An extension of the Media Lab building will open in the summer of 2009. A new cancer research facility, located next to the Koch Biology Building and across from the Broad Institute, is scheduled to open in December 2010. Also scheduled for completion in 2010 is the new home of the MIT Sloan School of Management, which will extend from Memorial Drive to Main Street and serve as an eastern gateway to the MIT campus.

Schools of MIT

School of Engineering(including 8 departments):
- Aeronautics and Astronautics Engineering
- Biological Engineering
- Chemical Engineering
- Civil and Environmental Engineering
- Electrical Engineering and Computer Science
- Materials Science and Engineering
- Mechanical Engineering
- Nuclear Science and Engineering

School of Humanities, Arts, and Social Sciences

Sloan School of Management

School of Science

School of Architecture and Planning

College of Health Sciences and Technology

Engineering research: The faculty, post-docs, graduate students, undergraduates, and other researchers that comprise the engineering community at MIT are singularly dedicated to the development of ideas, processes, materials, and devices that will improve the lives of people throughout the world. The School of Engineering's many departments, divisions, labs, and research centers collectively generate nearly $250 million in sponsored research every year—and they define the future of science and technology.

Engineering education: An engineering education at MIT prepares you to change the world.

Our emphasis on hands-on experience, project- and context-based learning, and leadership training make the School of Engineering's academic programs unique. And the deep interconnections between engineering curricula and the academic offerings from the Institute's other four schools ensure that what gets learned at MIT can be applied to the rest of the world.

Mechanical engineering: Mechanical engineering is one of the broadest and most versatile of the engineering professions. This is reflected in the portfolio of current activities in the department, one that has widened rapidly in the past decade. Today, our faculty are involved in projects ranging from the use of nano-engineering to develop thermoelectric energy converters to the use of active control of efficient combustion; from the design of miniature robots for extraterrestrial exploration to the creation of needle-free drug injectors; from the design of low-cost radio-frequency identification chips to the development of advanced numerical simulation techniques; from the development of unmanned underwater vehicles to the invention of cost-effective photovoltaic cells; from the desalination of seawater to the fabrication of 3D nanostructures out of 2D substrates.

Department of Mechanical Engineering at MIT

Mechanical engineering at MIT is nearly as old as MIT itself, and its impact on the Institute and on society itself is easily demonstrated by noting the alignment of the department's evolution with key events and technological advances in the world. The origins of Department of Mechanical Engineering trace back to the end of the American Civil War, in 1865. Its earliest areas of focus included extensive programs in power engineering and steam engines. By the mid-1870s, with the Industrial Revolution well underway in North America, the department became known officially. It innovated the use of lab subjects, giving students the opportunity to apply methodology to current engineering problems with hands-on lab work.

In 1874 mechanical engineering's first laboratory was built for direct application of current methodology to engineering problems. The education and research program of the new lab was applied in its approach and focused primarily on the steam engine. The specializations offered at the time reflected the industries of greatest growth, including marine engineering, locomotive engineering, textile engineering, and naval architecture. By the turn of the century through the advent of World War I, programs in steam turbine engineering, engine design, refrigeration, and aeronautical engineering set the stage for the technological advances to come.

Between World War I and World War II, automotive engineering was a very popular program in Department of Mechanical Engineering. The Sloan Automotive Laboratory, founded in 1929, became one of the world's leading automotive research centers. Post-World War II, the department's research emphasis gradually shifted from military applications (which continue to be an important component of the overall the department program in the present day) to "quality-of-life" applications, such as biomedical engineering, energy and environment, and human services.

By the mid 1970s, continuing to the present, research was concentrated within four major programs: biomedical engineering; energy and environment; human services, including transportation; and manufacturing, materials, and materials processing.

Today, Department of Mechanical Engineering attracts rich diversity and quantity of talented individuals, including 400 undergraduates, 500 graduate students, and about 75 faculty, many of whom are members of the National Academies and fellows of prestigious professional societies. Department of Mechanical Engineering conducts about $35 million worth of sponsored research annually, in a range of areas—such as mechanics, product design, energy, nano-engineering, ocean engineering, control, robotics, and bioengineering—that are diverse and yet also allow for rich collaboration both within the department and with other engineering and science disciplines at MIT and beyond.

These broad areas of focus on multidisciplinary research results in an exciting variety of innovative projects, including the use of active control to optimize combustion processes; the design of miniature robots for extraterrestrial exploration; the development of unmanned underwater vehicles; the prevention of material degradation in proton-exchange membrane fuel cells; the development of physiological models for the human liver; and the fabrication of 3D nanostructures out of 2D substrates.

Graduate education: MIT's graduate programs in mechanical engineering attract students with a variety of backgrounds, interests, and talents. We provide extensive opportunities for graduate students to engage in advanced research and collaborate with faculty and colleagues. Together, our community members push the boundaries of their professions, and grow profoundly as engineers, researchers, and innovators.

The mechanical engineering department provides opportunities for graduate work leading to the following degrees.

- Master of science in mechanical engineering

The SM in mechanical engineering is awarded based on the completion of advanced study and a major thesis. The thesis must be an original work of research, development, or design, performed under the supervision of a faculty or research staff member. Students usually spend as much time on thesis work as on coursework. This degree typically takes about one and one-half to two years to complete.

- Master of engineering in manufacturing

This twelve-month professional degree program prepares students to assume technical leadership in an existing or emerging manufacturing company. To earn this degree, students must complete a highly integrated set of projects that cover the process, product, system, and business aspects of manufacturing, as well as a group-based thesis project.

- Mechanical engineer's degree

This program provides an opportunity for further study beyond the master's level for those who wish to enter engineering practice rather than conduct further research. This degree emphasizes breadth of knowledge in mechanical engineering and its economic and social implications. It is quite distinct from the PhD program, which emphasizes depth and originality of research. The engineer's degree requires a broad program of advanced coursework and an applications-oriented thesis, and typically requires at least one year of study beyond the master's degree.

The PhD and ScD are the highest academic degrees offered. Doctoral degrees are awarded upon the completion of a program of advanced study in the student's principal area of interest, a minor program of study in a different field, and a thesis of significant original research, design, or development. Doctoral degrees are offered in all areas represented by the department's faculty.

B.2 参考译文

麻省理工学院

麻省理工学院（MIT）是美国一所著名的私立研究型大学，位于美国马萨诸塞州波士顿附近的的剑桥市，共有6个学院，32个系，历来重视科学技术方面的研究。麻省理工学院也是美国最早开设建筑、电气、卫生、造船、航海、航空、气象、核物理和人工智能等专业课程的学校。

MIT的四年全日制本科生教学计划重视学生的艺术与科学和专业技能的平衡发展；而研究生培养计划所涵盖的内容则更加系统和全面。该校教学水平得到了新英格兰学校协会[①]的认可。

按多种排名分类指标的评价，MIT都被列为美国和全世界的一流大学。其工学院的本科生及研究生的培养，从1994年《美国新闻与世界报道》公布的评价结果以来，一直被评为全美第一。1995年，美国国家研究协会的一份关于美国研究型大学的调查报告将MIT在美国的知名度排在第一，在文章引用和教师成果奖励方面排在第四；而2005年的一份研究则将MIT列为美国第四所世界各地莘莘学子最心仪向往的高校。

MIT的办学宗旨是引领知识，在科学、技术和其他方面培养学生，使其成为21世纪的人才，为国家和世界做出卓越的贡献。

---

① 新英格兰学校协会成立于1885年，是全美最古老的区域认证协会，其任务是建立和维持高标准的各级教育机构的教学质量。——编者注

MIT 在其办学章程被批准 4 年后的 1865 年开始招收第一批学生。这所名校的开办标志着对其创始人，著名的自然科学家 William Barton Rogers 为建立一所新型的、独立的、与日益工业化的美国发展密切相关的教育机构所做的不懈努力的最高嘉奖。Rogers 认为，专业技能的培养必须依靠教学和科研的结合，必须关注实际问题的解决。

今天，MIT 已成为一所世界级的教育机构。本着教学和科研为实践服务这一办学原则，MIT 始终遵循这一宗旨。MIT 是一所独立的、男女合校的、由私人资助的学校。下设的 6 个学院包含多个系、所和学位授予点，以及跨学科研究中心、实验室和培养计划，打破了传统的系际界线。

校 园 概 况

麻省理工学院占地 158 英亩，沿着查尔斯河，依傍剑桥市[①]的一侧，绵延约 1 英里。建于 1915 年的中央校区由一组互相连通的大楼组成，便于各院系之间的往来和交通。很多著名建筑师所设计的具有标志性的建筑，完成了其校园的规划扩展。

目前，在建的还有三座科研大楼。一座是媒体实验室扩建工程，将于 2009 年夏季竣工并对外开放；另一座是肿瘤研究中心，该建筑毗邻 Koch 生物学大楼，并从 Broad 研究所穿过，预期于 2010 年 12 月落成；第三座用于斯隆管理学院的新址，也将于 2010 年完工，它从 Memorial Drive 一直延伸到 Main Street，将成为 MIT 校园的东大门的标志。

院 系 设 置

工程学院（该院包括 8 个系）：
- 航空航天技术系
- 生物工程系
- 化学工程系
- 土木与环境工程系
- 电气工程与计算机科学系
- 材料科学与工程系
- 机械工程系
- 核科学与工程系

人文及社会科学学院
斯隆管理学院
自然科学学院
建筑与城市规划学院

[①] 剑桥市是毗邻波士顿西北部的一个小城市，人口仅 10 万，全球共有 780 人获得诺贝尔奖，其中有 130 人来自剑桥市，因而剑桥市驰名海外，成为世界各地莘莘学子心向神往、趋之若鹜的圣地。——编者注

健康科学技术学院[①]

工程研究　MIT 工程学科领域的教师、博士后研究人员、研究生、本科生及其他研究人员致力于新的观念、工艺方法、材料及装置的开发与研究，旨在不断改善人们的生活质量。工程学院的很多系所实验室和研究中心每年得到的科研经费达 2.5 亿美元，就是这些研究机构决定了未来的科学技术的发展。

工程教育　MIT 的工程教育会让你学会如何改变世界。其重点放在亲身实际经验的获得、课题和环境密切结合的学习，以及领导能力的培养上，从而使工程学院的培养计划别具特色。而该院工程类课程和学校其他 4 个学院的学科内容的密切结合足以保证在 MIT 学到的知识可以在世界任何其他地方都能得到应用。

机械工程　机械工程是应用范围最广、涵盖技能最多的工程专业之一。这可以从机械工程系当前各项教学科研活动中反映出来。在过去 10 年中，机械工程系得到了迅速的发展。今天，我系的教师所参与的项目范围很广，包括从纳米工程到热电能量转换器的开发到高效燃烧的主动控制；从用于外星探索的迷你型机器人设计到无针药物注射；从低成本射频识别芯片到先进数值仿真技术的开发；从无人水下交通工具的研制到成本效益高的光电池的发明；从海水淡化到在二维基片上制造三维纳米结构，等等。

机械工程系

机械工程系的历史和 MIT 的历史一样悠久，其对整个学校和社会的影响可以从该系的发展与世界上发生的大事和技术进步始终同步这一点充分体现出来。机械工程系的诞生可以追溯到美国 1865 年的南北战争时期。该系最早的学科重点包括动力工程和蒸汽机方面的系统的培养计划。到 19 世纪 70 年代中期，随着北美工业革命的发展，机械工程系已声名鹊起。该系创新利用了实验技术，给学生充分的机会将亲手操作的实验方法应用到解决当时的工程实际问题中。

1874 年，该系成立了机械工程实验室，将实验技术用于解决工程实际问题，并将教学与科研的重点放在蒸汽机技术上。那时的学科专业充分反映了当时工业界的伟大成就，包括航海工程、机车、纺织和造船等。从 20 世纪之交到第一次世界大战爆发，蒸汽轮机、发动机设计、制冷及航空工程等专业设置为美国后来的技术发展奠定了基础。

在两次世界大战之间的一段时间里，汽车工程是当时机械工程系最热门的专业。创办于 1929 年的斯隆汽车实验室后来成为世界一流的汽车研究中心。"二战"以后，该系的研究重点开始从军事应用（目前仍然是全系整体培养计划的重要构成部分）逐步转向提高"生活质量"方面的应用，如生物医学工程、能源与环境工程及公共事业等。

[①] 这里用 college 而非 school 只是命名时的习惯，大学里的学院采用 school 和 college 的都有，例如，北京大学下设学院的英文译名同时混用了这两种用法，物理学院称为 school of physics，而化工学院称为 college of chemistry；同济大学的各学院全部用 college，但医学院却用 medical school。——编者注

从 20 世纪 70 年代中期到现在，机械系的科学研究主要放在 4 个方面：生物工程，能源与环境工程，公共事业（包括运输），以及制造、材料和材料加工。

今天，机械工程系荟萃了各种各样的人才，包括 400 名本科生、500 名研究生和 75 名教师，不少教师是国家级研究院所和著名专业协会的成员。机械工程系每年的科研经费达 3500 万美元，其研究项目包括机械、产品设计、能源工程、纳米工程、海洋工程、控制技术、机器人和生物工程等多个领域，而且还可以和本系及 MIT 甚至其他院校的很多有关工程和科学学科进行充分的合作与交流。

这种广泛的研究领域十分重视各种多学科创新项目的研究，包括用主动控制优化燃烧过程、外星探索用的迷你型机器人设计、无人水下交通工具的研制、质子交换薄膜燃料电池材料的抗老化技术、人类肝组织生理模型的建立，以及在两维基片上制造三维纳米结构等。

研究生培养 MIT 机械工程系研究生的培养计划吸引了具有各种基础背景和个人兴趣的学生及其他各方面人才。我们为学生提供了各种充裕的机会从事先进技术的研究，以及和教师与同学之间的合作。大家在一起亲密合作，不断拓展其专业范围，为将来成长为优秀的工程师、研究人员和创新人才而努力奋斗。

机械工程系可以授予研究生以下几种学位。

- **机械工程理科硕士**
  机械工程的理科硕士需要完成高级研究项目和主修专业论文方可获得学位。毕业论文必须是在指导教师或有关研究人员指导下完成的原创性研究、开发或设计。学生要花费和课程学习同样多的时间撰写论文。完成学业大致要花一年半到两年的时间。

- **制造业的工程硕士**
  这是一种期限只有 12 个月的专业学位，主要针对那些打算在现有或新型制造业担任领导的学生设置的。要获得该学位，学生必须完成一整套完整的具体项目，包括工艺、产品、系统及制造经营方面的问题，论文撰写以组为单位集体完成。

- **机械工程师学位**
  这种学位是为那些已获得硕士学位，希望进入工程界而不打算从事进一步研究工作的人员继续学习而准备的。这种学位强调机械工程知识面的宽度及其经济和社会效益。它和强调研究工作的深度和原创性的 PhD 学位完全不同。工程师学位要求内容涉及较宽的先进课程的学习和面向应用的论文，一般需要在获得硕士学位后至少一年的时间来完成。

PhD 和 ScD（即哲学博士和科学博士）是最高级别的学位。要获得博士学位，学生必须完成主修学科和高级研究培养计划研究和另外一种其他领域的辅修专业，还要完成一篇有关原创性研究、设计或技术开发的重要论文。本系所涉及的各种专业领域均可授予博士学位。

附录 C 机床说明书翻译样本

Moore Nanotech® 350FG Ultra-Precision Freeform® Generator

Machine Features

- PC based CNC motion controller with Windows operating system and 1.0 nanometer programming resolution;
- Thermally insensitive linear scale feedback system with 0.034 nanometer resolution;
- Allows raster flycutting and/or grinding of freeform surfaces, linear diffractive surfaces, and prismatic optical structures;
- Allows increased swing capacity to 500mm dia. for off-axis and toric mponents;
- Box-way hydrostatic oil bearing slides with 300 mm of travel on Z, 350mm of travel on X, 140mm vertical travel on Y, and an adaptive air bearing counterbalance assembly on the vertical axis for optimal servo performance;
- Dual linear motors on Y-Axis;
- 10,000 rpm "eavy-duty" air bearing workspindle (with liquid cooling option) imbedded into the Y-Axis carriage to improve loop stiffness, reduce Abbe errors, and maintain symmetry;
- Options include hydrostatic rotary B-Axis, C-Axis positioning control of the workspindle, Fast Tool Servo system, grinding & micro-milling attachments, optical tool set station, spraymist coolant system, vacuum chuck, micro-height adjust tool holders, Aspheric Part Programming Software (APPS), on-machine measurement & Workpiece Error Compensation System (WECS), and air shower temperature control system.

Moore Nanotech® 350FG 超精密自由形面复合加工车床®说明书

机床特性

- 基于 PC 的数控加工运动控制器，Windows 操作系统，编程分辨率为 1.0 nm。
- 热稳定线性标尺反馈系统，分辨率为 0.034 nm。
- 飞刀车削或磨削加工自由形面、线性衍射表面和棱柱形光学器件。
- 偏摆加工直径为 500 mm 的非对称复曲面光学器件。
- 箱式静压油支持导轨，行程为：Z 向 300 mm，X 向 350 mm，Y（垂直）向 140 mm。Y 轴采用自适应空气轴承平衡装配，以保证优良的伺服控制性能。
- Y 轴配有双向直线电机。
- 转速为 10000 转/分的"重型"空气轴承支撑的液冷式工作主轴嵌入安装在 Y 轴的滑

座内，以提高环路刚度，减小阿贝误差[①]，维持结构对称性。
- 还有以下其他功能可任选：液压旋转 B 轴、工作主轴的 C 轴定位控制、快速刀具伺服控制系统、磨削和微铣削附件、光学对刀装置、喷雾冷却液、真空卡盘、刀架高度微调装置、非球面光学器件加工编程软件（APPS）、在线测量和工件误差补偿系统（WECS），以及空气簇射温度控制系统等。

表 C.1

General 各系统概述	Description 详细说明
System Configuration 系统配置	Ultra-Precision three, four or five axis CNC machining system for on-axis turning of aspheric and toroidal surfaces; slowslide-servo machining of optical freeform surfaces 超精密多轴（3，4，5 轴）慢速伺服数控系统，同轴车削加工光学自由曲面（非球面或环形表面）[②]
Workpiece Size 工件尺寸	500mm diameter × 300mm long 工件尺寸 Φ500 mm × 300 mm 长
Base Structure 基础结构	Monolithic cast epoxy-granite, with integral coolant troughs 基座为整体环氧树脂花岗岩浇注，集中式冷却液槽
Vibration Isolation 隔振	Optimally located three point passive air isolation system 三点优化定位结构的被动式空气隔振系统
Control System 控制系统	Delta Tau PC based CNC motion controller with 150 MHz DSP, operating in a Windows environment, with color flat panel touch screen display and PC-Anywhere remote diagnostics software with modem. 256 MB memory, AGP video, CD-RW / DVD Drive, and 80 GB hard drive 采用美国 Delta Tau 公司生产的基于 PC 的数控运动控制器，150 MHz 数字信号处理器，Windows 操作系统，彩色平板触摸屏显示器，带调制解调器的 PC-Anywhere 远程诊断软件，256 MB 内存。高级图形端口视频，可重写光盘驱动和 80 GB 硬盘驱动
Programming Resolution 编程分辨率	1 nanometer linear /0.00001° rotary 直线运动：1 nm；旋转运动：0.00001°
Machining Performance 加工性能	Material—High purity aluminum alloy. Form Accuracy (P-V)≤0.14 μm / 75 mm dia, 250 mm convex sphere. Surface Finish (Ra)≤3.0 nanometers 工件材料：高纯合金铝 形状精度≤0.14μm / 75 mm（峰-峰值），直径 250 mm 凸球面 表面粗糙度（Ra）≤3.0 nm

---

[①] 光学测量中由于镜片色散系数不同而造成的误差。——编者注
[②] 实际上，属于一种多轴（自由度）复合（车铣磨）金刚（金刚石刀具）车床，用于加工有色金属或其他硬质光学玻璃材料。——编者注

表 C.2

Workholding Spindle 工件夹紧主轴	Heavy Duty (Standard)　　重型(标准)
Type 轴承类型	Fully constrained Professional Instruments groove compensated air bearing 充分约束沟槽补偿式空气轴承
Liquid Cooling（optional） 液冷式（可选）	To maintain thermal stability and tool center repeatability, a closed loop chiller provides recirculating temperature controlled water to cooling channels located around the motor and bearing journals of the air bearing spindle. The chiller has an integral PID controller which maintains temperature control to ±0.5°F 为保证其热稳定性，采用闭环式冷却装置提供循环冷却水到主轴电动机和空气轴承轴颈部位的冷却流道。冷却装置用 PID 控制器，可控制温度在±0.5°F
Mounting 安装方式	Integrally mounted within the Y-axis carriage to increase loop stiffness and minimize thermal growth. Spindle cartridge resides in an athermal housing to further enhance thermal stability 整体安装在 Y 轴滑座中，以提高结构环路刚度和减少发热。主轴滑座装在不发热支座中，以进一步改善热稳定性
Speed Range 转速范围	50 to 10000 rpm, bi-directional 50—10000 rpm，双向
Load Capacity (Radial) 承载能力（径向）	36kg (80 lbs.) @ spindle nose 主轴端部的径向承载能力为 36 kg
Axial Stiffness 轴向刚度	140 N/μm (800000 lbs./in.)
Radial Stiffness (at nose) 径向刚度（轴端位置）	87 N/μm (500000 lbs./in.)
Drive System 驱动系统	Frameless，Brushless DC motor 无框架无电刷直流电动机
Motion Accuracy 运动精度	Axial≤25 nanometers; Radial≤25 nanometers 轴向≤25 nm；径向≤25 nm

表 C.3

Linear Axes 线性轴	X	Z	Y(vertical)　（垂直方向）
Travel 行程	350 mm	300 mm	140 mm
Drive System 驱动系统	Brushless DC Linear Motor 无刷直流直线电动机	Brushless DC Linear Motor 无刷直流直线电动机	Dual Brushless DC Linear Motor 双向无刷直流直线电动机
Feedback Type 反馈类型	Laser holographic linear scale (athermally mounted) 激光全息线性（无热安装）	Laser holographic linear scale (athermally mounted) 激光全息线性（无热安装）	Laser holographic linear scale (athermally mounted) 激光全息线性(无热安装)
Feedback Resolution 反馈分辨率	0.034 nanometer 0.034 nm	0.034 nanometer 0.034 nm	0.034 nanometer 0.034 nm
Feed Rate (maximum) 进给率（最大）	1400 mm/min	1400 mm/min	1400 mm/min

续表

Linear Axes 线性轴	X	Z	Y(vertical) （垂直方向）
Straightness in Critical Direction 敏感方向上的直线度	0.3 μm over full travel 全行程 0.3 μm	0.3 μm over full travel 全行程 0.3 μm	0.5 μm over full travel 0.3 μm (central)100 mm 全行程 0.5 μm；（中部）100 mm 范围内 0.3 μm
Hydrostatic Oil Supply 静压供油	Compact, low flow, low pressure system with closed loop servo control and pressure accumulator to minimize pump pulsation 结构紧凑，小流量，低压，闭环伺服控制，采用储压器减小泵压波动		

表 C.4

Optional Rotational Axes 可选旋转轴	B	C
Type 类型	Oil Hydrostatic 静压供油	Groove Compensated Air Bearing (liquid cooled) 沟槽补偿空气轴承（液冷）
Travel 旋转范围	360° (Bi-directional) 360°（双向）	360° (Bi-directional) 360°（双向）
Drive System 驱动系统	Brushless DC motor 无刷直流电动机	Brushless DC motor 无刷直流电动机
Axial Stiffness 轴向刚度	875 N/μm (5000000 lbs./in.)	140 N/μm (800000 lbs./in.)
Radial Stiffness (at nose) 径向刚度（主轴端部）	260 N/μm (1500000 lbs./in.)	87 N/μm (500000 lbs./in.)
Positioning Accuracy 定位精度	≤2.0 arc seconds (compensated) （补偿后）≤2 弧秒	≤±2.0 arc seconds (compensated) （补偿后）≤±2.0 弧秒
Feedback Resolution 反馈分辨率	0.02 arc seconds 0.02 弧秒	0.07 arc seconds 0.07 弧秒
Maximum Speed (Positioning Mode) 最大转速（定位模式）	50 rpm	1500 rpm
Motion Accuracy 运动精度	Axial:≤0.1 μm; Radial:≤0.1 μm 轴向≤0.1 μm；径向≤0.1 μm	Axial:≤0.025 μm; Radial≤0.025 μm 轴向≤0.025 μm；径向≤0.025 μm

表 C.5

Utility Requirements 设备安装要求条件	Air （压缩）空气参数	Electrical 电气条件	Floor Space 场地要求
For optimal cutting results, facility thermal stability should be held within±0.5℃ （±1.0°F） 为了保证最佳切削效果，该设备的热稳定性应保持在±0.5℃（±1.0°F）	7.5 to 9 bar (110—130psi) 425 liters/min; Dry to 10℃ pressure dew point and pre-filtered to 10μm 气压：7.5—9 bar（110—130 psi）；流量：425 升/分；干燥到压力露点为 10℃，过滤颗粒尺寸为 10 μm	11 kVA at the customer specified voltage from 220—480 VAC; 50/60 Hz; 3 Phase (26 kVA with optional oil hydrostatic grinder) 11 千伏安，电压 220—480 伏交流；频率 50/60 Hz；三相（若选用液压驱动磨削头，需用 26 千伏安供电）	1.93 m wide × 1.80 m deep × 2.06 m high Approx. 3,200 kg (Includes enclosure but not including pe-ripheral equipment and control pendant) 宽×深×高：1.93 m × 1.80 m × 2.06 m 总重约 3200 kg（包括包装但不包括辅助设备和悬挂式控制操纵盒）

附录D 校园常用词汇

拔尖人才 tip-top talent

班主任 class advisor

班长 class monitor

班委会学习委员 class committee member in charge of teaching/learning affairs

办学特色 school-running feature

保研到某大学 recommend for admission to the graduate school of … without taking matriculation examination

本硕（硕博）连读培养模式 continuous cultivating mode-from undergraduate to postgraduate (from master to doctoral students)

必修课 required (compulsory)course

毕业论文 thesis（本科生和硕士生），dissertation（博士生）

毕业答辩 oral defense for one's thesis (dissertation)

毕业实习 graduation field work

毕业设计 graduation project

毕业证 graduation certificate

闭卷考试 closed-book exam

博士后研究人员 post-doctoral fellow

博士基金 fund for doctoral dissertation

博士生导师 doctoral advisor (supervisor)

长江学者 Cheung Kong Scholar[①]

长江特聘教授 specially hired Cheung Kong professor

长江讲座教授 Cheung Kong chair professor

成绩单 transcript, school report, academic record

传道、授业、解惑 transmit wisdom, impart knowledge and resolve doubts

创建高等教育双一流 create double "first class" in higher education

创新团队 innovative team

大专 junior college

大专生 junior college student

---

① 长江学者奖励计划由教育部和李嘉诚基金会共同实施，Cheung Kong 是李氏长江集团的港式英文名称。

大专文凭　associate degree

多媒体教室　multimedia classroom

辅修课　minor course

复合型人才　inter-disciplinary talent

分数线　minimum passing score

个人简历　curriculum vitae (CV)，同 resume

高等院校　higher educational institutions, institutions of higher education

高考　national college entance examination

高考状元　*Gaokao* top scorer

国学　studies of Chinese culture

国家中长期教育改革和发展纲要　Outline of the National Program for Long- and Medium-term Educational Reform and Development

国家高层次人才特殊支持计划　National Plan for the Special Support of High-level Talents

基础研究　basic research（理科毕业论文类型）

基础课　basic course

技工学校　technical school

假文凭　fake diploma

价值观　values

奖学金　scholarship

交流学者　exchanged scholars

教学质量　teaching quality

教学改革　teaching reform

教育经费　education spending

教学评估　teaching quality evaluation

教务处　teaching affairs office, educational administration office

教书育人　impart knowledge and educate people

教育振兴行动计划　National Program of Action for Educational Vitalization

就业指导　employment instruction

考试成绩　exam scores

开卷考试　open-book exam

考研　take part in the post-graduate entrance examination

课件（教师用多媒体教学演示片）　courseware

课程设计　curricular project

课程表　curriculum schedule

课程设置　curriculum provision

科教兴国　rejuvenate our motherland through science and education

科学发展观　scientific outlook on development
理工科大学　university of science & engineering
录取线　admission bottom-line
录取通知书　admission notification
排名次序　ranking position
启发式教学　heuristic education
强化班　intensive training class
全国大学生XX挑战杯大奖赛（北京赛区）　National College Students Contest for "XX Challenge Cup"（Beijing venue）
勤工俭学　part-work and part-study
求职信　job application letter
人才流失　brain drain
人才交流　talent exchange
人才库　talent bank, brain bank
人才流动　mobility of talents
三好学生　"three good" student (an honorable title for outstanding students who are good in study, morality and health)
实验教学　laboratory teaching
师资　qualified teachers
首创精神　pioneering spirit
双学位　double degree
树立正确的世界观、人生观和价值观　foster a correct outlook on the world, life and values
素质教育　education for all-round development
授予学位　confer a ... degree on sb
双一流建设　doubl first-class construction
跳槽　job-hopping
通识教育（自由教育）liberal education
文娱委员　class committee member in charge of entertainment
校庆　anniversary of the founding of a school
（百年校庆 100th anniversary of the founding of ...）
校花　campus queen
校训　school motto
选修课　elective(optional) course
学霸　"overlord" of learning (extraordinarily excellent student)
学历　educational history
学费　tuition

学制　educational system

学位证　diploma

学位服　academic dress

学生会……部长　head of ... department of Student Union

学术报告　academic report

学术讲座　academic forum

学术交流　academic exchange

学生处　students' affairs division

学科建设　discipline construction, subject construction

学科带头人　pace-setter in scientific research, academic leader

研究生毕业证　graduate diploma

应届毕业生　this year's graduates

宣传委员　class committee member in charge of publicity

一等奖　first prize（注重奖金）, first-grade award（注重荣誉）

一等奖学金　first-rate scholarship

远程教育　distance education

因材施教　teach student according to their aptitude

应用基础研究　applied basic research（工科毕业论文类型）

优秀归国留学生（海归派）　excellent returned overseas Chinese students (scholars)

专业课　specialized course

自费留学　study abroad at one's own expense

自学考试　self-taught examination

在职研究生　on-the-job postgraduate

重点大学　key (prestigious, major, leading, banner, flagship) university

中专　secondary specialized school, polytechnic school

职校　vocational (training)school

招生办　admission office

助学金　stipend

振兴经济　revitalize the economy

振兴中华　rejuvenate the Chinese nation

附录 E 部分参考译文

第 2 单元 工程材料（Ⅱ）

2.1 前言

在第 1 单元中，我们简单介绍了金属材料。本单元我们将对其他三类常用的工程材料，即陶瓷、复合材料和聚合物进行一个大致的介绍。

2.2 陶瓷

陶瓷是金属和非金属元素的化合物。由于各元素之间有多种可能的结合状态，所以目前用于消费品及工业产品的陶瓷种类非常多。比如，陶瓷由于具有极好的绝缘和高温强度，已用做汽车火花塞很多年。它们在工模具、热机及汽车零部件如排气管衬套、涂层活塞和汽缸套方面的应用越来越多。

陶瓷可分为两大类：

1. 传统陶瓷，如白色陶瓷、砖瓦、下水管道及砂轮。
2. 工业陶瓷，又称为工程陶瓷、精细陶瓷或先进陶瓷。它们在涡轮机、汽车、飞机、热交换器、半导体、密封件、假肢和刀具等方面应用很广。

陶瓷可以以单晶或多晶的形式存在，其晶粒大小对强度和性能影响很大，晶粒越小，它们的强度及韧性也就越高。

氧化物陶瓷有氧化铝和氧化锆两种。

氧化铝又称为刚玉和金刚砂，它是应用最广的氧化物陶瓷。它们可以是纯三氧化二铝，或者是与其他氧化物组成的混合物。这类陶瓷的硬度高且具有一定的强度。虽然自然界就存在有氧化铝，但它一般都和其他一些含量不等的杂质混合在一起，性质很不均匀，所以其性能不是很可靠。现在，氧化铝、碳化硅和其他一些陶瓷几乎全都是人工合成方法制备而成的，故可对其质量进行控制。

氧化锆（分子式 ZrO_2，白色）具有很好的韧性、抗热冲击性、耐磨性、低的导热性和摩擦系数。部分稳定氧化锆具有较高的强度和韧性，比氧化锆性能更可靠。它是通过对氧化锆进行钙、钇和镁的氧化物掺杂制成的一种新型陶瓷。这类材料的典型应用包括金属零件的热挤压模具、用于磨削的锆砂粒、航空件涂层的分散剂、汽车车身表面的底漆和面层涂料，以及用于食品柔性包装上的光面图片。

其他陶瓷还有以下几种：

碳化物。典型实例有用做刀具和模具材料的碳化钨和碳化钛，以及用于磨料（砂轮）的碳化硅。

1. 碳化钨陶瓷由碳化钨颗粒与黏结剂钴组成。黏结剂用量对材料性能影响很大。黏结剂钴用量增加将提高其韧性，而硬度、强度和耐磨性却会降低。
2. 碳化钛用镍和钼作为黏结剂，韧性不如碳化钨。
3. 碳化硅具有很好的耐磨、耐热冲击性和耐腐蚀性。其摩擦系数小，在高温下仍具有一定强度，适合用做热机的高温零部件，还可用做磨料。

氮化物。另外一种十分重要的陶瓷是氮化物，特别是立方氮化硼、氮化钛和氮化硅。

1. 立方氮化硼是一种仅次于金刚石的材料，在刀具和砂轮磨料上具有特殊用途。它在自然界并不存在，是在 20 世纪 70 年代首次由人工合成得到的，其合成方法类似人造金刚石。
2. 氮化钛广泛用于刀具涂层，凭借较低的摩擦特性，它可改善刀具寿命。
3. 氮化硅具有很好的抗高温蠕热变性能、低热膨胀性、高的导热系数，因此可抵抗热冲击。这种陶瓷材料适合用做高温结构材料，如汽车发动机和涡轮机零件、凸轮推杆滚柱、轴承零件、喷砂嘴及造纸工业的零部件。

金属陶瓷。金属陶瓷是 20 世纪 60 年代出现的一种由陶瓷相和金属相构成的混合物，又称为黑色陶瓷和热压陶瓷。它们将陶瓷的抗高温氧化性，以及金属材料的韧性、抗热振性和延展性结合在一起。它的一种典型应用是用做刀具材料，其构成成分为 70%氧化铝和 30%氮化钛。

其他金属陶瓷含有各种氧化物、碳化物和氮化物，主要用于高温环境，如喷气发动机的喷嘴和飞机制动器。它也可看作是一种复合材料，可以用粉末冶金技术将不同成分的陶瓷与金属粘连起来。

由于陶瓷在高温下仍能保持较高的强度和刚性，因而在高温工作环境下得到广泛应用。它们的耐磨性使得这些材料很适合用做汽缸套、衬套，密封件和轴承。陶瓷的高温工作性能也意味着将提高汽车燃料燃烧效率，减少排放。目前，内燃机的燃烧效率只有 30%左右，但若采用陶瓷零件，则其性能将至少改善 30%。

目前，工作温度高达 1000°C（1830°F）的全陶瓷热机的材料与技术开发已在进行之中。然而，由于性能不稳定、缺乏足够的韧性、对轴承和热工作部件的润滑困难，并考虑到发动机的尺寸精度，要求最后进行机加工和精整加工，从而缺乏低成本生产的近净成形的结构陶瓷（如氮化硅和碳化硅）的能力，故这种全陶瓷结构的热力发动机的发展进程最终比预期要慢。

生物陶瓷。由于其强度和化学稳定性（惰性），陶瓷常常用做生物材料来代替人体关节、假肢及牙科用料。而且，陶瓷材料制作的人工植入体可以具有多孔结构（如同多孔钛合金材料一样），便于骨细胞的长入，使其新生骨和陶瓷植入体具有一定的结合强度。常用生物陶瓷有氧化铝、氮化硅及各种硅的化合物。

2.3 聚合物

聚合物具有很大的分子量，由聚合物制成的消费品或工业用品包括食品和饮料容器、包装材料、信号标识、纺织品、医疗用品、泡沫材料、涂料、安全防护材料、玩具、家电、镜头、齿轮、电子电器产品、汽车车身及其部件等。聚合物材料具有以下各种独特性能：

1. 耐腐蚀耐化学性能；
2. 低导电导热性；
3. 低密度；
4. 高比强度，尤其是经过增强处理后；
5. 降噪性；
6. 颜色与透明度可任意选择。

现代塑料技术开始于 20 世纪 20 年代，当时制作聚合物的原材料是从煤和石油产品中提炼出来的。乙烯就是最早采用的原材料，后来成为制备聚乙烯的基本材料。乙烯是乙炔和氢的反应生成物，而乙炔又是由焦炭和甲烷制成的。常用聚合物材料包括聚丙烯、聚氯乙烯、聚碳酸酯等都是用类似工艺制成的，它们都称为合成有机高分子材料。

热塑性材料。如果我们对某种聚合物材料加热后冷却，它会恢复到原来的硬度和强度的话，换言之，如果这样的过程是可逆的，那么它们就是热塑性的。典型的这种材料有丙烯酸、纤维素、尼龙、聚乙烯、聚氯乙烯。

热固性材料。然而，如果聚合物的长链分子交联成三维结构，就会成为大分子。这种大分子是一种强有力的共价键结构，这时的聚合物就成为热固性聚合物。因为聚合反应时形成了网状结构，所以聚合后的零件形状也就永久保留不变了。此时的固化（交联）反应和热塑性材料不同，不再是可逆的了。如果温度继续升到一定高度，热固性聚合物就会烧损、分解以至碳化。和热塑性材料相比起来，热固性材料一般具有较好的机械、热、化学性能，较高的电阻及尺寸稳定性。典型的热固性材料有酚醛树脂，是苯酚与甲醛的反应生成物。由这种材料制成的常见产品有炊具用品的手柄和灯具的开关与插座等。

可降解塑料。塑料废品约占城市固体垃圾的 10%。三分之一的塑料产品使用后将废弃，如塑料瓶、塑料包装和垃圾袋。随着塑料产品的使用越来越广泛，对环境问题越来越关注及垃圾填埋处越来越稀缺，人们正在开发研制能完全降解的塑料产品。大多数塑料产品传统上都是由合成聚合物制成的，而这些材料又是由不可再生的自然资源制备而成，它们不能降解，很难回收。聚合物的生物降解性是指环境中的微生物物种（例如，土壤和水中的微生物）在合适的条件下可以降解部分或全部聚合物材料，而不生成有毒的副产品。降解后的终端产品就是二氧化碳和水。由于可降解塑料还含有各种各样的其他成分，可被看成复合材料，所以这些塑料中只有一部分是真正能完全降解的。

2.4 复合材料

近年来，材料科学发展的一个重大成就是复合材料。事实上，复合材料现在已成为最重要的工程材料之一，这是因为它们和传统材料相比具有很多优越的性能。复合材料是由两种以上化学性能完全不同且互不溶混的相构成的。它们的性质与结构特征要优于单独作用时其各个组成部分。

例如，大家都知道，塑料的机械性能要比金属和合金差，特别是它们的强度、刚性和抗蠕变性能是无法与金属相比的。但这些性能可以通过各种形式的包埋增强（例如，加入玻璃或石墨纤维）来得到改善，从而生产出增强塑料。金属和陶瓷也可以采用加入塑料或纤维的方式来提高其性能。这种材料被称为金属基和陶瓷基复合材料。

增强塑料的应用。增强塑料最早（1907 年）的用途是用于耐酸容器，它是由酚醛树脂和石棉纤维构成的。后来环氧树脂在 20 世纪 30 年代被首次用做复合材料的基材。20 世纪 40 年代，玻璃纤维被用来制作小型船体。复合材料的真正发展是在 20 世纪 70 年代，这种材料现在被称为先进复合材料。它们主要用在军用和民用飞机、火箭零部件、直升机螺旋桨、汽车车身、板簧、传动轴、管道、梯子、压力容器、运动器材、头盔、船体及各式各样的结构件上。玻璃纤维和碳纤维增强的混杂型塑料目前主要用于高温环境，它们可在 300°C（550°F）下连续工作。波音 777 的全部构件总重的 9%是复合材料制成的，相当于过去波音运输机复合材料的 3 倍。它们的底板横梁、仪表盘及大多数垂直和水平尾翼都是由复合材料制成的。由于减轻了飞机重量，增强塑料可为飞机节省燃油约 2%。

金属基复合材料。金属基复合材料比高分子聚合物用做基材的优势是具有更高的弹性模量和高温性能，而且韧性和延展性也更好。不足之处是密度稍大，加工困难一些。这类基体材料常用的有铝、铝锂合金、钛、铜、镁及超级合金。纤维材料有石墨、氧化铝、碳化硅、硼、钼和钨等。非金属纤维材料的弹性模量一般为 200～400 GPa，抗拉强度为 2000～3000 MPa。硼纤维增强的铝基复合材料由于它们具有很高的比刚度，重量轻及导热性能好，因此已经用做航天飞机轨道飞行器的结构管件支架。

陶瓷基复合材料。陶瓷基复合材料是工程材料的又一重大成就，因为它们具有很好的抗高温及抗腐蚀性能。一般来说，陶瓷结实而坚硬，抗高温，但韧性较差。能在 1700°C（3000°F）温度下保持其较高强度的基体材料有碳化硅、氮化硅和氧化铝。现在各种用来改善陶瓷基复合材料的机械性能，尤其是提高它们的韧性的技术已在进行研究和开发。实际应用有喷气发动机、汽车发动机、深海采矿设备、压力容器、结构件、刀具及金属材料的挤压和拉伸模具等。

超材料。 超材料指那些由人工设计其结构并具有非同一般甚至极不寻常特性的材料，其性能远远超过了它们的组成成分或其它自然或人造材料。Meta 这个单词来源于希腊语，意思是超越或超级。超材料具有一些超越自然材料的特殊性质。它们的这些性质来源于它们的特殊结构而不是组成它们的材料成分所贡献的结果。光学、声学和热学领域的超材料的重大进展激发了人们对机械超材料的研究。机械超材料表现出了许多不同寻常的性质比方负的可压缩性，负泊松比，负质量密度和弹性模量以及特殊的弹性动力学性能。

超材料的类型：一般说来，超材料可以分为机械超材料、电磁超材料、和声学超材料等。在过去的 10 来年，人们已对超材料开展了深入详细的研究。由于超材料大大扩展了材料在设计波控器件方面的利用空间，目前又对其相应的共振机理、动态均质化理论和微结构设计方面进行了广泛的探索。超材料的迅速发展给它们带来了广阔的应用前景，也许最不可思议的就是采用工程化手段专门制成的超材料，它们具有特殊的物理性质，可开发这类材料专门应用于某些特别的技术领域。其中最吸引人的例子就是隐身斗篷和超级透镜。人工构造的超材料已经成为光学和电磁学领域制作具有复杂的空间域和频率域性能的有用工具。可重构和可调谐的超材料的最新进展有可能进一步扩展到制作具有实际功能的超材料器件和特殊的亚波长器件。这类超材料器件除了表现出某些目前自然界没有电磁响应外，它们还有可能在一些小型器件的多功能应用领域表现出更加优良的性能。

第3单元　铸　　造

3.1　前言

将金属加工成可以使用的产品有各种不同的方法，如铸、锻、焊和机械加工等。人们习惯上将铸、锻、焊称为成形加工。这种工艺通常要采用模具和某些工艺装备迫使熔融金属或者毛坯成形为最终零件。而机械加工（包括车削、铣削、磨削和钻削等）是采用刀具或其他物理、化学手段[①]将毛坯表面不需要的部分以切屑的形式去除掉。铸造是最古老的成形工艺之一，它是将熔融金属注入模具型腔，使其冷却后成形为铸件。铸造工艺早在公元前4000年左右就用来制作装饰品、铜箭头及其他各种物件。铸造可以整体生产制作复杂的零件，包括有内部空腔的产品如发动机缸体等。几乎所有的金属材料都可以铸成任意形状的零件，常常只需少许的最后精加工，因此铸造被认为是最重要的净成形制造技术之一。

3.2　砂型铸造

传统的金属铸造方法就是砂型铸造，已经有几千年的历史。简单说来，砂型铸造由以下几个工艺步骤构成：（1）将与铸件最终形状完全一样的母模置入砂中做成模腔；（2）安上浇注系统；（3）注入熔融金属；（4）使金属凝固成形；（5）打破砂模；（6）取出铸件。

砂模主要构成如下：

1. 由砂箱包围的铸型。两分式铸型由上、下型箱构成，接缝处称为分型面。如果铸型由两个以上部分构成，中间部分称为中箱。
2. 浇口盆或浇口杯，熔融金属由此注入。
3. 直浇口，引导熔融金属朝下注入。
4. 横浇口系统，引导金属从直浇口注入型腔，内浇口就是金属最后流入铸型型腔的进口处。

---

① 例如，电火花加工、化学铣、激光加工等。——编者注

5. 冒口，当铸件凝固收缩时用做补充金属液。图 3.1[②]中显示了两种不同的冒口，即明冒口和暗冒口。
6. 铸芯，实际上就是用砂制成的嵌块。它们置入铸模后就形成空腔部分或形成铸件的某个内表面。有时，这些芯子也可用来形成铸件的表面特征如字符图形，或者比较深的表面凹腔等。
7. 通气孔，用来排放当熔融金属与砂型或芯子接触后产生的气体或熔融金属进入型腔时排挤出的空气。

母模是用来使铸砂形成注入熔融金属的型腔的，它们一般用木材、塑料或金属制成。其选材主要取决于铸件的大小、形状、尺寸精度、铸件数量及制模工艺等因素。

3.3 熔模铸造

熔模铸造工艺又称为失蜡铸造，起源于公元前 4000 至公元前 3000 年。它所用的母模可用蜡或塑料（如聚苯乙烯），也可用快速原型制作。

熔模铸造的操作工序如图 3.2 所示。母模是通过将熔融的蜡或塑料注入金属模具中成形的。然后，将其浸入由耐火材料如微细石英粉和包含有水、硅酸乙酯和酸的黏合剂组成的浆料中。当这层初始涂挂材料干燥后，蜡模再经多次涂覆耐火浆料使涂层达到一定厚度。熔模铸造之所以称为 investment casting，主要是因为母模要被耐火材料所包裹，invest（覆盖）就是 cover（涂覆）的意思。蜡模强度较低，在制模过程中一定要小心。不过，它和塑料还不一样，（除非是热塑性塑料）蜡可以回收重用。当蜡模熔化后，便留下模腔以便熔融金属的注入，形成最终的铸件，即使很复杂的形状也能制作得具有很高的精度。特别是难于加工的金属，尤其适合于采用熔模铸造工艺来制作金属零件。

3.4 发泡模铸造

发泡模铸造采用（膨胀发泡的）聚苯乙烯做母模，这种聚苯乙烯被熔融金属气化后在模型中留下一个型腔来生产铸件。因此，该工艺又称为气化模或消失模铸造，其商业名称又称为实型铸造。过去称为"聚苯乙烯胀型工艺"，现已成为最重要的黑色金属和有色金属铸造方法，尤其在汽车工业中应用很广。

该法采用可膨胀的聚苯乙烯颗粒作为原料，加上 5%～8%戊烷（一种可挥发的碳氢化合物），置入预热好的铝制模具中。聚苯乙烯发泡膨胀后充填型腔，剩余的热量使聚苯乙烯颗粒熔化聚结，待模具冷却后就可打开取出聚苯乙烯模型。如果模型比较复杂，则可用分块制作后用热熔粘接的办法制作母模。

聚苯乙烯母模表面涂挂上一层水基耐火浆料，干燥处理后置入砂箱。砂箱中充填松散的细砂，将母模包裹（图 3.3），也可混入某种黏合剂使其具有一定强度。型砂还可采用各种方

---

② 参考译文中的图请参见本书正文部分。

式捣实。然后，不用取出聚苯乙烯母模，直接注入熔融金属液，使聚苯乙烯立即烧蚀气化并同时充填型腔，取代曾被聚苯乙烯母模占据的部分。聚苯乙烯被热解（解聚反应）后的分解物通过周围型砂逸散排出。

3.5 离心铸造

顾名思义，这种铸造工艺是利用旋转产生的惯性力将熔融金属液注入型腔制作铸件的。该工艺诞生于 19 世纪早期。一般圆筒状零件如管道、枪筒、道路灯柱都可采用这种工艺铸造（图 3.4），它们是将熔融金属注入旋转的模具中制成的。一般，旋转轴呈水平布置。模具采用钢、铸铁或石墨制成，内壁可涂上耐火衬料，以提高模具寿命。

模具表面也可做成各种形状，使得管道外形可以包括矩形和多边形等。但铸件内壁仍然保持为圆柱形，因为这样熔融金属可以在离心力作用下均匀分布注入。不过，由于密度不同，较轻的材料如熔渣、夹杂物及耐火衬料等，可能会聚积在铸件内表面。直径为 13 mm（0.5 in.）~3 m（10 ft），长度达 16 m（50 ft）的铸件都可用该法铸造。铸件壁厚范围在 6 mm 到 125 mm（0.25~5 in.）之间。因为产生的离心力很大，所以可以铸造厚壁件。该法可以用来生产具有很高尺寸精度及外壁具有精细图形或花纹的优质铸件。除管道外，其他典型铸件还包括套筒、发动机缸体衬套、带有凸缘或不带凸缘的轴承圈。

3.6 铸件检验

有好几种方法可以用来检查铸件的质量或任何可能存在的缺陷。首先，铸件可用目测或光学方法检查表面质量。表皮下面或内部缺陷可用各种非破坏性检测技术检查。在破坏性试验中，试样从铸件的各个断面取得并进行强度、延展性及其他机械性能的检测，从而确定产生缩松和其他任何缺陷的位置所在。

铸件（如阀、泵和管件等）的耐压性通常可采用密封铸件开口后注入加压的水、油和空气的方法来检查（不过有一点要千万注意，空气是可压缩的，在试验中一定要注意由于铸件的某处缺陷所导致的突然爆裂的发生）。对于密封性要求极高的情况，可用压缩氦或专门注入带有气味并装有传感器（嗅味器）的办法来试验，然后在压力保持下测试其密封性能。不合格或有缺陷的铸件则重熔再铸。

由于铸件缺陷会影响生产的经济效益，因此有必要对铸件缺陷和产生原因进行研究。在铸造过程中的各个生产阶段，从模型制备到从铸型取出铸件的各个环节都对保证产品质量有重要影响。

第 4 单元　锻造与模具

4.1 前言

锻造是一种通过各种模具或其他工具将工件进行压力加工成形的工艺。它是最古老的一种金属加工技术，可以追溯到公元前 4000 年甚至公元前 8000 年。最早是用来制作首饰、钱币及各式各样的其他器具，一般采用石制工具对金属进行锤击加工。

简单的锻造就是用手锤和铁砧进行加工，通常由铁匠来完成。但大多数锻件需要一套模具及像压力机和锻锤这样的设备来完成。与通常用来制作连续型板材、带材及各种不同断面结构的轧制不同，锻造是生产离散型零件的。

典型的锻件有螺钉、铆钉、连杆、涡轮轴、齿轮、手工工具，以及机械、飞机、铁路和各式各样其他运输设备所用的构件等。

由于锻造可以对金属的流动和晶粒结构进行控制，因此锻件具有很好的强度和韧性，能可靠地用于承受较大应力和要求严格的工作环境。锻造可以在室温（冷锻）或高温（温锻或热锻）下进行。①

由于材料强度较高，冷锻加工需要的压力更大，而且工具材料应在室温下具有很好的延展性。冷锻件表面光洁度和精度较高。热锻需要的力较小，但产品的尺寸精度和表面光洁度比冷锻要差一些。

锻件一般需要后续精加工如热处理，来改善其性能。然后，再经机械加工获得精确的最终尺寸。如果采用精密锻造，则这些步骤就可省去，这是净成形或近净成形工艺的发展趋势，它将大大减少操作步骤，从而降低最终产品的制造成本。

4.2 自由锻

自由锻是最简单的一种锻造方法。虽然大多数自由锻件的质量在 15～500 kg (30～1000 lb)，但质量达 300 t 的锻件也可生产出来。锻件尺寸可以从很小的零件到 23 m 长的轴（船用螺旋桨）。

自由锻的操作过程是将工件放在两个平模之间，在压力作用下减小其厚度，故又称为墩粗或平模锻。自由锻模具的表面可以有简单的凹腔来生产相对较简单的锻件。由于锻造过程中毛坯体积一般保持不变，所以减小锻件高度就会增加其直径（如锻打圆盘类零件时）。

4.3 锻模与闭式模锻造

在锻模锻造时，锻件毛坯搁在两个成形模具之间锻造，从而获得模腔的形状（见图 4.1）。此时，我们会发现有坯料向外流动形成飞边。锻件的飞边对金属材料在模锻过程中的流动有重要的影响。飞边较薄，冷却较快，造成摩擦阻力，使模腔中的金属材料由于无法继续向外流动而受到很大的压力，有利于它在模腔中的充填。

坯料放在下模中，当上模下压时，它将逐渐发生形状的改变，一根连杆的锻造过程如图 4.2(a)所示。（锻件毛坯的）预成形如压肩和卡压［见图 4.2(b)，(c)］是对毛坯材料在不同区域进行分配，如我们在制作糕点时首先要将面团揉捏成初始的形状一样。卡压是使坯料从

---

① 这里有一点要说明，金属的再结晶温度是其熔点（热力学温度 K）的 0.4～0.6，在该温度以上锻造属于热锻，因此铅即使在室温下加工也算是热锻。——编者注

某处向两端分配，而压肩则是使坯料集中在某个局部区域。然后，锻坯被预锻模初轧成连杆大致的形状。最后一道工序是在锻模中进行精锻，以获得零件的最终形状，图 4.1 和图 4.2(a)所示的操作也称为闭式模锻造。然而，在真正的闭式模锻或无飞边锻造中，不会有飞边产生，坯料将完全充满模具型腔。精确的控制坯料体积与适当的模具设计对于获得具有令人满意的尺寸和公差的闭式模锻件是至关重要的。若毛坯尺寸不够，则会使模腔充填不满；反之，若尺寸过大，则会导致太大的压力而使模具永久性损坏或"卡死"。

4.4 精密锻造

出于经济方面的考虑，今天的锻造工艺正趋向越来越精密，大大减少了后续精加工工序的数量。我们将成形零件接近最终尺寸的加工工艺称为净成形或近净成形。在这种加工方式中，锻件没有什么多余的材料，即使有也可通过随后的处理（去毛刺或磨削）去除掉。

精密锻造使用特殊模具，生产的锻件精度比锻模锻造更高，后续机加工也少得多。这种工艺需要生产能力更大的设备，因为要获得零件的细节（形状尺寸），锻造力载荷会更大。由于铝和镁锻造时需要的载荷和温度相对较低，因此它们特别适合于精密锻造，而且，模具磨损也小，锻件表面光洁度较高。不过，钢和钛有时也可进行精密锻造。典型的精锻工件有齿轮、连杆、机架及涡轮机叶片。

精压。精压是一种典型的闭式模锻成形工艺，常用于硬币、徽章和首饰的制作；小片的金属坯料在完全闭合的模腔中成形。为了获得锻件的细节形状与尺寸，锻造压力比毛坯材料强度高 5—6 倍。某些零件甚至需要经过好几道精压工序。此时不能采用润滑剂，因为润滑剂也可能被封闭在模腔中，由于它们是不可压缩的，从而会阻碍模具表面细节的完全充填。精压工艺也用于锻件和其他零件的最终加工，以改善其光洁度和保证零件的尺寸精度。

4.5 模具制造方法

模具制造有各种各样的方法，可以是单一的，也可以由多种方法组合而成。这些方法包括铸、锻、机械加工、磨削及电加工和电化学加工仿形制模等。为了获得较高的硬度和耐磨性，模具通常需要进行热处理。必要时还需要通过精磨和抛光来改善工件的表面轮廓形状与光洁度，具体操作可用人工方式和可编程的工业机器人。

模具制造方法的选择主要取决于模具操作条件和尺寸形状等因素。当然，成本也是要考虑的，因为工模具的制造往往要花费可观的成本。例如，一套汽车车身覆盖件模具要花费 200 万美元。即使一些小型简单模具，也需要花费几百美元。而另一方面，由于一个模具可以生产大量零件，每个零件的模具成本只占零件整个制造成本的一小部分。

锻模一般可以分为阳模与阴模，也可以按尺寸大小来分类。小模具的面积为 1000～10000mm^2，而较大的模具面积可达 1m^2 以上，比方用于汽车覆盖件冲压加工的模具。

不同形状尺寸的模具可以用钢、铸铁或有色金属制成。所用的工艺包括砂型铸造（对于几吨重的模具而言）到壳型铸造（小模具）。常用的模具材料有钢、模具钢、高速钢甚至碳化物。铸铁一般用来制作大件的模具，因为它们的强度与韧性及成分、晶粒尺寸与性能都可以比较容易地进行控制和改善。

不过，模具最常用的制造方法还是用锻坯进行机加工，如铣、车、磨及电加工和电化学加工。典型的热作模具是用数控铣加工的。当然，如果要加工很硬的或经过热处理的高强度耐磨模具材料，机加工也是十分困难的。

这些操作很费时，因此非常规加工方法也比较广泛，尤其是加工中小型模具。这些加工方法加工速度快、经济，模具无须以后的精加工。例如，金刚石拉丝模就是用涂有金刚石粉的很细的高速旋转的细针（用油作为润滑剂）来加工模具上的小孔的。

为了提高硬度、耐磨性和强度，模具钢常常要进行热处理。而热处理不当则是模具损坏的最常见的原因之一。由于热处理过程中微结构发生变化及不均匀的热循环作用会使模具发生变形，因此这种情况尤其受模具表面状况和化学成分的影响。

热处理之后，模具要经过最后的精加工，如磨削、抛光、电化学加工等来获得较好的表面光洁度和尺寸精度。对于磨削，如果操作不当会因为过热造成模具表面损伤，在模具表面会产生有害的残余拉伸应力，从而降低其疲劳寿命。另外，模具表面的刮擦也会提高表面应力。同样，常用的制模工艺如电火花加工，假如工艺参数不小心控制的话，也会造成表面损伤和裂纹。

第6单元　特种加工工艺

6.1　前言

传统加工如车削、铣削和磨削等，是利用机械能将金属从工件上剪切掉，以加工成孔或去除余料。特种加工是指这样一组加工工艺，它们通过各种涉及机械能、热能、电能、化学能或及其组合形式的技术，而不使用传统加工所必需的尖锐刀具来去除工件表面的多余材料。

传统加工如车削、钻削、刨削、铣削和磨削，都难以加工特别硬的或脆性材料。采用传统方法加工这类材料就意味着对时间和能量要求有所增加，从而导致成本增加。在某些情况下，传统加工可能行不通。由于在加工过程中会产生残余应力，传统加工方法还会造成刀具磨损，损坏产品质量。基于以下各种特殊理由，特种加工工艺或称为先进制造工艺，可以应用于采用传统加工方法不可行，不令人满意或者不经济的场合：

1. 对于传统加工难以夹紧的非常硬的脆性材料；
2. 当工件柔性很大或很薄时；
3. 当零件的形状过于复杂时；
4. 要求加工出的零件没有毛刺或残余应力。

传统加工可以定义为利用机械（运动）能的加工方法，而特种加工利用其他形式的能量，主要有如下三种形式：

1. 热能；
2. 化学能；

3. 电能。

为了满足额外的加工条件的要求，已经开发出了几类特种加工工艺。恰当地使用这些加工工艺可以获得很多优于传统加工工艺的好处。常见的特种加工工艺描述如下。

6.2 电火花加工

电火花加工是使用最为广泛的特种加工工艺之一。相比于利用不同刀具进行金属切削和磨削的常规加工，电火花加工更为吸引人之处在于它利用工件和电极间的一系列重复产生的（脉冲）离散电火花所产生的热电作用，从工件表面通过电腐蚀去除掉多余的材料。

传统加工工艺依靠硬质刀具或磨料去除较软的材料，而特种加工工艺如电火花加工，则是利用电火花或热能来电蚀除余料，以获得所需的零件形状。因此，材料的硬度不再是电火花加工中的关键因素。

电火花加工是利用存储在电容器组中的电能（一般为 50 V/10 A 量级）在工具电极（阴极）和工件电极（阳极）之间的微小间隙间进行放电来去除材料的。如图 6.1 所示，在 EDM 操作初始，在工具电极和工件电极间施以高电压。这个高电压可以在工具电极和工件电极窄缝间的绝缘电介质中产生电场。这就会使悬浮在电介质中的导电粒子聚集在电场最强处。当工具电极和工件电极之间的势能差足够大时，电介质被击穿，从而在电介质流体中会产生瞬时电火花，将少量材料从工件表面蚀除掉。每次电火花所蚀除掉的材料量通常在 $10^{-6}\sim10^{-5}$ mm^3 范围内。电极之间的间隙只有千分之几英寸，通过伺服机构驱动和控制工具电极的进给使该值保持常量。

6.3 化学加工

化学加工是众所周知的特种加工工艺之一，它将工件浸入化学溶液通过腐蚀溶解作用将多余材料从工件上去除掉。该工艺是最古老的特种加工工艺，主要用于凹腔和轮廓加工，以及从具有高的比刚度的零件表面去除余料。化学加工广泛用于为多种工业应用（如微机电系统和半导体行业）制造微型零件。

化学加工将工件浸入到化学试剂或蚀刻剂中，位于工件选区的材料通过发生在金属溶蚀或化学溶解过程中的电化学微电池作用被去除掉。而被称为保护层的特殊涂层所保护下的区域中的材料则不会被去除。不过，这种受控的化学溶解过程同时也会蚀除掉所有暴露在表面的材料，尽管去除的渗透率只有 0.0025～0.1 mm/min。该工艺采用如下几种形式：凹坑加工、轮廓加工和整体金属去除的化学铣，在薄板上进行蚀刻的化学造型，在微电子领域中利用光敏抗蚀剂完成蚀刻的光化学加工（PCM），采用弱化学试剂进行抛光或去毛刺的电化学抛光，以及利用单一化学活性喷射的化学喷射加工等。如图 6.2(a)所示的化学加工示意图，由于蚀刻剂沿垂直和水平方向开始蚀除材料，钻蚀（又称为淘蚀）量进一步加大，如图 6.2(b)所示的保护体边缘下面的区域。在化学造型中最典型的公差范围可保持在材料厚度的±10%左右。为了提高生产率，在化学加工前，毛坯件材料应采用其他工艺方法（如机械加工）进行预成形加工。湿度和温度也会导致工件尺寸发生改变。通过改变蚀刻剂和控制工件加工环境，这种尺寸改变可以减小到最小。

6.4 电化学加工

电化学金属去除方法是一种最有用的特种加工方法。尽管利用电解作用作为金属加工手段是近代的事,但其基本原理是法拉第定律。利用阳极溶解,电化学加工可以去除具有导电性质工件的材料,而无须机械能和热能。这个加工过程一般用于在高强度材料上加工复杂形腔和形状,特别是在航空工业中如涡轮机叶片、喷气发动机零件和喷嘴,以及在汽车业(发动机铸件和齿轮)和医疗卫生业中。最近,还将电化学加工应用于电子工业的微加工中。

图 6.3 所示的是一个去除金属的电化学加工过程,其基本原理与电镀原理正好相反。在电化学加工过程中,从阳极(工件)上蚀除下的粒子移向阴极(加工工具)。金属的去除由一个合适形状的工具电极来完成,最终加工出来的零件具有给定的形状、尺寸和表面光洁度。在电化学加工过程中,工具电极的形状逐渐被转移或复制到工件上。型腔的形状正好是与工具相匹配的阴模的形状。为了获得电化学过程形状复制的高精度和高的材料去除率,需要采用高的电流密度(范围为 10~100 A/cm^2)和低电压(范围为 8~30 V)。通过将工具电极向去除工件表面材料的方向进给,加工间隙要维持在 0.1 mm 范围内,而进给率一般为 0.1~20 mm/min。泵压后的电解液以高达 5~50 m/s 的速度通过间隙,将溶解后的材料、气体和热量带走。因此,当被蚀除的材料还没来得及附着到工具电极上时,就被电解液带走了。

作为一种非机械式金属去除加工方法,ECM 可以以高切削量加工任何导电材料,而无须考虑材料的机械性能。特别是在电化学加工中,材料去除率与被加工件的硬度、韧性及其他特性无关。对于利用机械方法难于加工的材料,电化学加工可以保证将该材料加工出复杂形状的零件,这就不需要制造出硬度高于工件的刀具,而且也不会造成刀具磨损。由于工具和工件间没有接触,电化学加工是加工薄壁、易变形零件及表面容易破裂的脆性材料的首选。

6.5 激光束加工

LASER 是英文 Light Amplification by Stimulated Emission of Radiation 各单词头一个字母所组成的缩写词。虽然激光在某些场合可用来作为放大器,但它的主要用途是光激射振荡器,或者是作为将电能转换为具有高度准直性光束的换能器。由激光发射出的光能具有不同于其他光源的特点:光谱纯度好、方向性好及具有高的聚焦功率密度。

激光加工就是利用激光和靶材间的相互作用去除材料。简而言之,这些加工工艺包括激光打孔、激光切割、激光焊接、激光刻槽和激光刻划等。

激光加工(见图 6.4)可以实现局部的非接触加工,而且对加工件几乎没有作用力。这种加工工艺去除材料的量很小,可以说是"逐个原子"地去除材料。由于这个原因,激光切割所产生的切口非常窄。激光打孔深度可以控制到每个激光脉冲不超过 1 μm,且可以根据加工要求很灵活地留下非常浅的永久性标记。采用这种方法可以节省材料,这对于贵重材料或微加工中的精密结构而言非常重要。可以精确控制材料去除率使得激光加工成为微制造和微电子技术中非常重要的加工方法。厚度小于 20 mm 的板材的激光切割加工速度快、柔性好、质量高。另外，通过套孔加工还可有效实现大孔及复杂轮廓的加工。

激光加工中的热影响区相对较窄，其重铸层只有几微米。基于此，激光加工的变形可以不予考虑。激光加工适用于任何可以很好地吸收激光辐射的材料，而传统加工工艺必须针对不同硬度和耐磨性的材料选择合适的刀具。采用传统加工方法，非常难以加工硬脆材料如陶瓷等，而激光加工是解决此类问题的最好选择。

激光切割的边缘光滑且洁净，无须进一步处理。激光打孔可以加工用其他方法难以加工的高深径比的孔。激光加工可以加工出高质量的小盲孔、槽、表面微造型和表面印痕。激光技术正处于高速发展期，激光加工也如此。激光加工不会挂渣，没有毛边，可以精确控制几何精度。随着激光技术的快速发展，激光加工的质量正在稳步提高。

6.6 超声加工

超声加工为日益增长的对脆性材料如单晶体、玻璃、多晶陶瓷材料的加工需求及不断提高的工件复杂形状和轮廓加工提供了解决手段。这种加工过程不产生热量、无化学反应，加工出的零件在微结构、化学和物理特性方面都不发生变化，可以获得无应力加工表面。因此，超声加工被广泛应用于传统加工难以切削的硬脆材料。在超声加工中，实际切削由液体中的悬浮磨粒或者旋转的电镀金刚石工具来完成。超声加工的变型有静止（传统）超声加工和旋转超声加工。

传统的超声加工是利用作为小振幅振动的工具与工件之间不断循环的含有磨粒的浆料的磨蚀作用去除材料的。成形工具本身并不磨蚀工件，是受激振动的工具通过激励浆料液流中的磨料不断缓和而均匀地磨损工件，从而在工件表面留下与工具相对应的精确形状。音极工具振动的均匀性使超声加工只能完成小型零件的加工，特别是直径小于 100 mm 的零件。

超声加工系统包括音极组件、超声发生器、磨料供给系统及操作人员的控制。音极是暴露在超声波振动中的一小块金属或工具，它将振动能传给某个元件，从而激励浆料中的磨粒。超声加工系统的示意图如图 6.5 所示。音极/工具组件由换能器、变幅杆和音极组成。换能器将电脉冲转换成垂直冲程，垂直冲程再传给变幅杆进行放大或压抑。调节后的冲程再传给音极/工具组件。此时，工具表面的振动幅值为 20～50 μm。工具的振幅通常与所使用的磨粒直径大致相等。

磨料供给系统将由水和磨粒组成的浆料送至切削区，磨粒通常为碳化硅或碳化硼。另外，除了提供磨粒进行切削外，浆料还可对音极进行冷却，并将切削区的磨粒和切屑带走。

第 20 单元　现代制造技术的发展

20.1　前言

尽管不能很精确地判定制造业的起源年代，但一般可以追溯到公元前 5000 至公元前 4000

年。它要早于有记录的历史，因为原始山洞或者岩石上的一些印记和绘画都要凭借一些刷子或其他用"颜料"来做标记的工具，或在岩石上凿槽的方法去完成，必须有相应的工具来完成这些工作。不同用途的产品的制造，起始于用木头、陶土、石头或者铁等材料所做的物品的生产。经历了若干世纪，最初通过铸造和捶击来改变产品形状的材料与工序已通过采用新材料和更加复杂的操作方法，正在逐步向不断提高生产率和产品质量的方向发展。

现在制造业已发展成一个生产产业，它通过将原材料转换成产品而获取利润。而利润是来自于产品开发、质量、可靠性、定价、公众形象、生产率及团队工作等。制造业的主要目的是将原材料制成高品质货物，这些货物在市场上有需求价值且能以具有竞争力的价格进行销售。通过良好的管理技巧来生产这些货物，并使用诸如资金、劳力、材料、设备和能源等涉及很多活动和操作的资源。制造可以定义为一系列有关的活动与操作，包括设计、选材、计划、生产、质量保证、管理，以及离散消费品和耐用产品的市场营销等。这些活动和操作中的相互作用构成了一个总的制造系统。该系统是人力资源，机器、工具和设备资源，财政资源，以及用来实现一组特定功能的方法的有机集合。许多制造工艺都是根据规范标准将一组新的材料和形状参数转换成作为输出量的产品尺寸、结构和性能，从而实现某个制造系统的主要目标。

人们在不断寻找新的方法，以生产出能提高他们的生活水平，满足他们的基本需要乃至控制他们的生活环境的一些设备。这些改进了的方法产生了对一些设备的大量需求，这些设备能以比以前更快的速度生产出单件成本更低但质量却更好的产品。

20.2 机械化

用于生产的一些简单机械的构建和应用始于约 1670 年的欧洲。这些发展使得产品生产从家庭转移到工厂，标志着工业革命的起源。机械化意味着产品由在家中手工制造转移到工厂由机器生产。

机械化也创造了大批量生产系统，它对机器提出了能高精度复制零件的要求。这就导致了对于更精确的测量工具、不断改进的测量技巧及能帮助制造者制造互换件的标准的需求。大批量生产带来的主要结果就是刚性自动化生产机构和自动化生产线。自动化生产线是由一些能在更短的生产时间内更快地生产产品的制造设备组成。

继刚性自动化之后，出现了一些具有简单的自动控制功能的机床。这种类型的控制器通过调节被控变量或系统而自动运行。控制技术的进步开辟了自动化的新纪元，称之为可编程自动化。

20.3 可编程自动化

为了适应产品的变化，可编程自动化应运而生。一项自动化领域的新技术——数字控制（NC）于 1952 年前后被开发出来。数控是基于电子计算机的原理，通过数字、字母和符号来控制生产过程的可编程序自动化的一种形式。计算机技术的提高使数控延伸到直接数控（DNC）、计算机数控（CNC）、图形数控（GNC）和语言数控（VNC）。

数控引起了生产离散型金属零件的革命。数控的成功导致了许多相关技术的延伸，如自适性控制（AC）和工业机器人。自适应控制将加工过程中的工件材料硬度、切削宽度或深度等因素作为变量来确定合适的加工速度和进给量。自适应控制指的是测量特定的输出过程变量并用它们来控制机床速度和进给量的一个控制系统。自适应控制加工系统中典型的工艺变量包括主轴的弯曲变形或力、力矩、切削温度、振幅和功率。在 20 世纪 70 年代后期，工业机器人开始在制造业中起主要作用。最初机器人主要用于材料搬运，然而今天的机器人技术已发展到能在制造中完成许多复杂高级的操作任务。

20.4 计算机辅助制造

计算机在制造系统中起着日趋重要的作用。由于计算机具有可以快速接收和处理大量数据的能力，使得这种（依靠计算机发展起来的）系统方法成为现代制造系统中必不可少的控制与处理手段。计算机在制造中的应用已经很成熟。在制造生产中，计算机的典型应用归类为计算机辅助制造（CAM）。它是构筑在诸如数控（NC）系统、自适应控制（AC）系统、机器人系统、车辆自动导航系统（AGVS）和柔性制造系统(FMS)等基础之上的。

CAM 是计算机技术通过直接或间接的与生产公司之间物质和人力资源的计算机接口，在制造生产设备的规划、管理、控制和运行中的一种有效应用。计算机在 CAM 系统中起着重要的作用。它们将工艺数据整合成共用的数据库。数据库管理的概念被应用于 CAM 操作中，来加快数据存取，确保所有使用者围绕共同的设计任务工作。在 CAM 的定义下，计算机应用包括诸如库存控制、进度计划、设备监控和信息管理等系统。这些应用主要用于对制造数据的转换、解释和跟踪。

20.5 柔性

在现代制造中，柔性是一个很重要特性，它意味着一个制造系统具有各种功能，并有很强的适应性，而且还能够应付较大批量的生产过程。柔性制造系统的多功能体现在能生产不同的零件。适应性表现为系统可迅速地改造，以适应生产完全不同的零件。柔性可在竞争激烈的国际市场中决定成败。

制造中生产率与柔性之间有一个折中。自动化生产线产量高但柔性不够。另一方面是能提供最大的柔性但产量较低的独立 CNC 设备，柔性制造试图在柔性与生产率之间获得最佳平衡。

自动化生产线能够以高生产率生产大量的零部件，这种生产线安装时间长，但能生产大量的相同部件。它的主要缺点是即使零件有很小的设计更改，都会造成整个生产线的关闭和重新装备。这是一个致命的弱点，意味着如果不对连续生产线进行代价昂贵和耗费时间的关闭与重新装备，它就无法生产不同的零件，即使该零件属于同一零件族也不行。

CNC 机床通常用来生产小批量零件，它们在设计上稍有区别。因为它们能够很快重新编程，以适应很小的甚至重大的设计更改，所以这些设备是达到此目的的理想选择。然而，作为独立的设备，它们不能以高生产率生产大批量的零件。

柔性制造系统（FMS）由一组计算机数控机床和一个为其服务的由计算机控制的具有换

刀能力的自动化物料搬运系统组成。由于这样的换刀能力和计算机控制，该系统能连续重新构建以生产不同种类的零件。这就是它为什么称之为柔性制造系统的缘故。

FMS 与独立的 CNC 机床相比，能以更高的生产率处理更大批量的生产。柔性制造能力的特别显著之处是大多数制造环境需要中等生产率来生产中等批量的产品，并且以足够的柔性来迅速重构以生产另一类产品。柔性制造填补了这一在制造业中长期存在的空白。

柔性制造朝着完全集成制造的目标跨出了重要的一步，因为它涉及自动化生产过程的集成。在柔性制造中，自动化制造设备（如车床、磨床、钻床）和自动化物料搬运系统，通过计算机网络共享即时通信。这是一种小规模的集成。

柔性制造通过集成几个自动化制造理念朝着完全集成制造的目标迈出了重要的一步：
1. 单个机床的计算机数控（CNC）；
2. 制造系统的分布式数控（DNC）；
3. 自动化物料搬运系统；
4. 成组技术（零件族）。

当这些自动化处理过程、设备和理念带入到一个集成系统时，FMS 便诞生了。在 FMS 中，人和计算机起着主要作用。但工人数量远少于人工操作的制造系统。然而，人仍然在 FMS 的操作中起着重要的作用。

20.6 再制造

每年产生的无害废物中超过 60% 来自制造业，日益严厉的法规要求产品和制造加工过程对环境减小影响。例如，制造商责任立法要求制造商回收使用过的产品，以减少垃圾填埋地。这样的压力，伴随着全球工业激烈的竞争，对公司产品设计态度的改变提出了挑战。公司必须设计长寿命产品，且在其寿命周期结束时便于回收产品材料，并且必须考虑处理废旧产品的商业潜质，从而利用产品零部件的剩余价值。

再制造是这样一个过程，它将使用过的产品在其功能上"整旧如新"，此过程不仅是可以获利的，而且比常规制造过程对环境危害作用小，因为它能减小垃圾填埋坑的占用面积和在生产过程中原材料、能源及具有专门技术的劳力的消耗。再制造的关键障碍包括消费者的接受程度、再制造工具和技巧的稀缺及现行许多产品不具备再制造性。它源自于再制造知识的缺乏，包括对其模棱两可的定义。修理、更新和再制造这样一些术语通常一起使用，因此消费者不能确定再制造产品的质量，对购买它们持谨慎态度。而且，设计师在他们的工作中，缺乏相关知识去考虑再制造这样的寿命终结问题，因为传统上设计只关心产品的功能性和在环境问题上付出的代价。另外，对生产企业组织改造涉及的许多问题需要进行调查，如技术创新、企业组织与社会发展水平的创新及这些改造的时间表之间的相互作用。需要对制造管理中的决策过程，特别是在改变主动性或改造中的决策过程进行评价。再制造代表了大多数制造商决策过程的根本变化。从历史角度来看，废物管理方法的选择是基于对可量化、可度量的成本和经济利益的纯经济分析。但这个方法却忽略了在决策过程中所存在的许多影响选择合适技术的定性因素。另一方面，再制造是一个复杂的、多学科的、多功能的作业方法，它潜在决定了工业中可利用的大量的废物最小量化技术，例如，产品的变化、生产过程中原材料的变化、操作实践中的变化及产品回收技术等。它需要若干设计方案和基于数据的活动之间的相互协调，例如，环境影响分析、数据和数据库管理以及设计优化。决策要受不可预测性、风险性和不确定性的影响，这一点已形成共识。例如，某种技术的影响能精确计算吗？如何识别和计算不同方面的风险（如经济、社会和环境）？为什么在一个企业生产组织中成功的某一实践，在另一企业中却是失败的？

第 21 单元　信息技术与制造业

21.1　前言

什么是 IT？IT（信息技术）是指利用电子计算机和现代通信手段获取、识别、传递、存储、处理、显示和分配信息的技术。在过去的几十年内，IT 在制造业中的广泛应用使得通信和信息交换变得更加经济有效。IT 在制造业中的应用范围从简单的机加工到生产规划和控制支持。从数控技术的诞生直到今天更加完善的通信与信息交换，信息技术给我们带来的好处有协同设计、加工中心、制造单元和柔性制造、远程与网络化制造等。

计算机将制造带入了信息时代。一个高度发展的、充满竞争的世界正在要求制造业开始满足更多的需求，并使其自身采用先进高端的技术进行装备。例如，为了适应竞争，一个公司会满足某些多少有点相互矛盾的需求：既要生产多样化和高质量的产品，又要提高生产率和降低价格。在努力满足这些要求的过程中，公司需要一个采用先进高端的工具，一个能够对顾客的需求做出快速反应，而且从制造资源中获得最大收益的工具。这个工具就是计算机。

21.2　计算机集成制造

要成为一个具有"超高质量和超高生产能力"的工厂，需要对一个非常复杂的系统进行集成。这只有通过采用计算机对机械制造的所有组成部分——设计、加工、装配、质量保证、管理和材料装卸及输送等环节进行集成才能够完成。

例如，在产品设计期间，交互式的计算机辅助设计系统使得完成绘图和分析所需要的时间比原来缩短，而且精确程度得到了很大的提高。此外，样机的试验与评价程序进一步加快了设计过程。

在制订制造计划时，计算机辅助工艺规划可以从数以千计的工序和加工过程中选择最好的加工方案。

在车间里，分布式智能以微处理器控制的机器这种形式来操纵自动装卸料设备和收集关于当前车间状态的信息。

但是这些各自独立的改革还远远不够。我们所需要的是由一个通用软件从始端到终端进行控制的全部自动化的系统。

"计算机集成制造"（CIM）是用于工厂全自动化的一个术语，它使所有功能都受计算机控制而且用数字化信息将它们紧密联系起来。CIM 包含了许多其他的先进制造技术，如 CNC、CAD/CAM、机器人及准时（JIT）交货。计算机集成可提供广泛、及时和精确的信息，可以改进各部门之间的交流与沟通，实施更严格的控制，从而提高整个系统的整体质量与效率。

随着计算机时代的到来，制造业的发展已经完成了整整一个循环。设计已经从使用诸如计算尺、三角板、铅笔、比例尺和橡皮擦的手工工艺演变成一种称为计算机辅助设计（CAD）的自动化工艺。工艺规划从采用计划表、框图和图表等手工处理演变成一种称为计算机辅助工艺规划（CAPP）的自动化过程。生产从手工操纵机床发展成一种被称为计算机辅助制造（CAM）的自动化过程。这些制造的各个环节与部门经过多年的演变和发展，已成为独立自动化岛。这些孤岛和制造业的其他自动化部门通过计算机网络被连接在一起。

现代制造业包含了将原材料制成产品、将成品送入市场及产品的现场维护等所有必要的活动与过程，它们包括如下：

1. 某产品需求的识别；
2. 设计满足需求的产品；
3. 获得生产产品的原材料；
4. 运用合适的加工工艺将原材料转化成产品；
5. 输送产品到市场；
6. 现场维护产品，以保证其良好的性能。

借助于 CIM，不仅使不同的生产环节自动化了，而且这些自动化小岛都连接在一起或者说集成在一起。集成化意味着一个系统能够提供完全的和及时的信息共享。在现代制造业中，由计算机完成集成的工作。CIM 是包含从原材料到成品再到市场的所有环节的总集成。

将来的完全 CIM 系统包含五个主要环节：产品设计、生产规划、生产控制、生产设备和生产过程。它具备三个方面的潜力。第一点，只有计算机能让制造业具有两种强大的、从未有过的性能，即

1. 在线可变程序（柔性）自动化；
2. 在线实时最优化。

第二点，计算机不仅能为制造的硬件部分（制造的机器和设备），而且能为制造的软件部分（信息流通，数据库等）做以上事情。第三点，也是最重要的，不仅在制造活动中的各个环节，而且在整个制造系统中，计算机都有能力完成上述工作。因此，计算机具有巨大的潜能来集成整个系统，产生所谓的计算机集成制造系统。

21.3 企业资源规划（ERP）系统

现在 ERP 系统整合了整个企业的资源规划和业务处理，包括人力资源、项目管理、产品设计、材料和生产能力规划。错误信息和数据冗余的剔除、业务单元的界面标准化、全局存取和安全问题之间的矛盾及业务流程的准确建模等，所有这些都是一个成功实施的 ERP 的组成部分。ERP 系统能提供最优化处理的能力，这就导致不同制造工序的省钱省时。显示的案例包含从简单的优化问题、车间调度和生产规划问题到当今日复杂的决策问题和实时生产控

制。此外，通常来说，ERP 系统不仅能提供供应链管理（SCM）的解决方案，而且还可提供与其他外部信息技术系统以集成方式进行交互的界面。SCM 解决方案处理制造企业的当前发展趋势，通过将企业运作与商业合作伙伴的运作相结合，使企业的通信和协作能力达到最大化。

21.4　计算机辅助系统 CAx

由于和 ERP 系统相结合，计算机辅助系统通过拓展设计者的创造性，对新产品和工艺概念进行早期试验，最小化对样机的需求，同时考虑拆卸和回收方法，并对能源和材料优化使用，从而大大提高了生产率。最新的 CAx 方案集成了设计（CAD）、工艺规划（CAPP）、工程（CAE）、质量控制（CAQC），提供设计和型号选择、完全可重构的工艺过程规划、机加工 NC 代码、有限元分析、静力学和动力学分析、公差规范的一致性及几何特性（包括了材料的质地和机械性能）的完全仿真。对于看似遥远的集成阶段，已出现了新的方法，如工艺规划和调度，因此允许对可用资源进行更有效的协调。而与之互补的，在数字环境中的协同设计是另一个新兴的很有前景的研究开发领域。共享虚拟环境的开发使得分散的操作者能够共享数据并使之可视化，在通过网络进行产品设计和工艺过程设计中进行真实交互和决策。CAx 系统辅助工程团队解决出现在生产现场的问题，通过使用这些新技术可以有效实现材料和能源的优化利用。

ERP 和 CAx 的集成化发展形成了一个达到最新技术发展水平的产品寿命周期管理系统（PLM），从该系统可以得到各种各样的数据管理任务，包括数据储存、工作流程、寿命周期、产品结构，以及检查和改变管理任务。另一方面，产品数据管理系统（PDM）宣称能够集成和管理大量的生产应用、信息类型及加工过程。这一过程从设计、制造到终端用户的支持对一个产品给予了确切的定义。制造组织具有将企业的各种业务功能和它们的各个部门及上述这些新系统（PLM 和 PDM）集成为一个企业数据库的能力，这样就使得它们具有统一的企业眼光。上述这些系统是基于数字化工厂/制造理念的，根据这样的理念，在开始生产之前，要将生产数据管理系统和仿真技术联合起来使用，以优化制造过程。原则上，通过开发数字化制造，制造企业希望达到以下几个目的：

1. 缩短产品开发周期；
2. 制造工艺过程的早期确认；
3. 更快地提高生产能力；
4. 加快上市时间；
5. 降低制造成本；
6. 提高产品质量；
7. 扩大产品知识宣传；
8. 减少差错；
9. 增加柔性。

除了核心硬件工程的优势外，随着通过信息社会（如网络）而不断增强的竞争，还会有一些经济方面的优势。因此可以认为，这些新技术的积极采用能弥补巨大的材料和能源需求。总之，为了耗费更少而生产更多，生产的柔性化和自动化正变得越来越重要。

参 考 文 献

[1] WRIGHT P K. 21st Century manufacturing[M]. New Jersey: Prentice Hall, 2001.

[2] KALPAKJIAN S. Manufacturing engineering and technology[M]. New Jersey: Prentice Hall, 2001.

[3] DOYLE L E, KEYSER C A. Manufacturing processes and materials for engineers[M]. New Jersey: Prentice Hall, 1985.

[4] EDWARD G H. Jig and fixture design[M]. New York: Delmar Publishers Inc., 1980.

[5] ROBERT Q. Computer numerical control[M]. New Jersey: Prentice Hall, 2004.

[6] BERND B, JAY F T. A thermal model for high speed motorized spindles[J]. Machine Tools and Manufacture, 1999, 39(2): 1345-1366.

[7] IJOMAH W L, MCMAHON C A, HAMMOND G P, etc. development of design for remanufacturing guidelines to support sustainable manufacturing[J]. Robotics and Computer-Integrated Manufacturing, 2007, 23(8): 712-719.

[8] DORF R C, KUISIAK A. Handbook of design, manufacturing and automation[M]. New York: John Wiley & Sons, 1994.

[9] GARCIA E J. The evolution of robotics research[J]. Robotics & Automation Magazine, 2007, 14(1): 90-103.

[10] LYNCH M. Computer numerical control[M]. New York: McGraw-Hill, 1994.